The Kingdom of Oil

Also by Ray Vicker

THOSE SWISS MONEY MEN

The Kingdom
of Oil

THE MIDDLE EAST:
ITS PEOPLE AND ITS POWER

Ray Vicker

CHARLES SCRIBNER'S SONS

NEW YORK

Library of Congress Cataloging in Publication Data

Vicker, Ray.
  The Kingdom of oil.

  1. Petroleum industry and trade—Near East.
2. Near East—Social conditions. I. Title.
HD9576.N36V52   338.2'7'2820956   73–19654
ISBN 0–684–13728–3

CONTENTS

The Kingdom of Oil

I

America's Oil Allies

There was the hustle and excitement of great events in Abu Dhabi's Ministry of Petroleum Building along the Persian Gulf shore. Robed officials hurried in and out, stepping deftly over construction debris before the entrance. A photographer's flash popped. Headlights of arriving automobiles swept like searchlights across that unlit lobby every time a vehicle turned into the parking lot, where a cement mixer partially blocked the way.

The Ministry Building was so new that steps to the first floor still were under construction. Access to the upstairs was only by the elevator. The offices above had the wall-to-wall-carpeted elegance that Arab oil barons favor. Golden drapes hung by windows. Italian marble coffee tables held cut-glass ashtrays. The chandeliers might have come from a Venetian palace.

In truth, the atmosphere of unlimited wealth was not feigned. Abu Dhabi is a small, desert sheikhdom in the United Arab Emirates that figuratively floats on oil. Moreover, the oil is a low-sulfur, high-quality product that admirably fits into the programs of ecologists who want to reduce the noxious fumes of energy in the making. The petroleum is in strong demand; and the demand pumps money into Abu Dhabi in a rushing flow.

The lights of Abu Dhabi's new seaport a mile down the promenade blinked and glowed in the evening haze. Auto traffic on the drive consisted largely of Japanese-made Toyotas and Datsuns serving as taxis for the Bahrainis, Afghans, Baluchis, Pakistanis, Indians, Americans, Britishers, and other foreigners who have rushed to this new Coronado on the Gulf just as the forty-niners poured into California in another age. Not for nothing has oil been described as black gold.

Mana Saeed Al-Otaiba, Abu Dhabi's Minister for Oil, had announced a press conference for 6:00 P.M. Arab journalists for papers in Cairo, Kuwait, Bahrain, Beirut, and elsewhere were gathering in the minister's reception office. I was the only American among the dozen and a half reporters, feeling somewhat self-conscious in that I also was the only non-Arab.

I was there because I knew that this would be an historic conference, one that would mark the end of an era in the international oil industry. It also would mark the beginning of a new, troubled era for an oil-short United States. America definitely does not float on oil, even though its economy is fueled by it.

Within a few years, the U.S. will be importing over half the oil it consumes. This is something to face now. Those who won't think of the future may be condemned to live in a wretched present.

In 1972, the United States obtained about 26 percent of its crude oil and 12 percent of its total energy requirements from foreign sources. By 1985, this country is likely to be importing 10 to 15 million barrels of oil per day, or as much as 55 percent of the oil that America would be consuming. Much of these imports can be supplied only by the Middle East, or perhaps one should say "by Middle Easterners." Economic language often describes a situation in an abstract, nonpersonalized manner, as if people were not involved in the massive movements dictated by geo-economic factors.

Perhaps it is a measure of the changes underway in international oil that a meeting in Riyadh, Saudi Arabia, being announced in small Abu Dhabi could have worldwide implications. Much oil policy, today, is made in producer states, not in the board rooms of major companies. Arabs and Iranians have 85 percent of the non-socialist world's crude oil reserves that are outside the United States and Canada. They account for 90 percent of the oil going into world markets.

Not long ago, the major oil companies held all the aces in the international energy game. These companies are Exxon, Texaco,

Gulf, Mobiloil, Standard Oil Company of California, British Petroleum Company (BP) and Shell. The first five are American, BP is a British company half-owned by the British government, and Shell is an Anglo-Dutch combine. Whenever any crisis developed in the petroleum industry, some oil official was apt to say smugly. "The Arabs can't drink their oil." This was supposed to conjure a picture of unhappy Arabs grouped around unsold barrels of oil like merchants with overstocked warehouses.

The major companies had downstream operations, networks of filling stations, refineries, tank farms, transport fleets, and other facilities for taking the crude black liquid from the ground and refining it into something useful, then delivering it to customers willing to pay the prices.

Today, the companies still have the downstream systems, but they receive less and less American oil to put into them. Meanwhile, they have also lost control of the Middle Eastern oil that once flowed at their command.

Today, you don't hear much talk about the drinkability or non-drinkability of Arab oil. Nobody wants the Arabs to even suggest that perhaps they will hold their oil awhile as they consider what to do with it. That oil is needed too badly in the Western world and in Japan. Within the span of a few short years the power has shifted. Oil-producing nations, not companies, are now in a position to say how much of their oil will be pumped from the ground, where it will be sold, and how profits will be distributed.

Give a country an opportunity to fix its share of any profits and invariably it will find reasons for boosting its percentage of the take. Nothing is more sacrosanct (in its own mind anyway) than the rights of a specific country when that nation's vital interests are at stake. Economic or military ruthlessness seems justified in those circumstances. Wars are fought on that plane. Economic squeezes are introduced without qualms. Now Middle Eastern oil-producing nations are in a position to increase their share of petroleum revenues at the expense of the oil companies and their customers.

This was the reason why an announcement in Abu Dhabi could hold the attention of the international oil industry. When everybody is listening even a small drum is enough.

It was also the reason why Arab oil countries and Iran could dictate an increase in their revenues from about 88 cents a barrel in 1969 to $1.50 in 1972, and to about $1.75 in 1973. Their total revenues should triple again by 1975, then more than double again between that time and 1982.

This shift of power within the oil industry has profound implications for every American who drives an automobile, heats his home with oil or natural gas, or consumes electricity at home. His way of life may be changing. He may find that his living standards are in hostage to those Middle Easterners, most of them Arab. They hold their fingers on the triggers of the world's oil nozzles.

It is easy to visualize a possible scenario which might take place in many of thousands of American communities some time in the period ahead. It is a sparkling day in the spring, the sort of time when the motorist enjoys being behind the wheel of an automobile, preferably in the country where the land is greening in the strengthing sun. Motorist Jones is at the wheel of his car, with several chores scheduled in a leisurely day. Almost mechanically, he turns into a Texaco filling station to fill the car's tank, as he has done every Saturday morning for years.

At the station an attendant is flushing the concrete with a water hose, the sheen of the water sparkling in the sunshine. His manner provides no warning that the pump tank is empty.

"Sorry," he says. "We're all out."

Further down the street the attendant at the Exxon station gives the same response. He is locking the station as if he might have been expecting a long gasoline drought. Almost gloatingly, he says: "You won't find a loose gallon in town. Everybody is sold out."

Only then does Jones realize that perhaps this isn't one of those periodic shortages which had been occurring of late. The scarcity of traffic this morning should have warned him, as might the radio if he had depended upon it rather than taped music for travel sound.

That remark of the attendant echoed in his head. Not a loose gallon in town. And already he was below a quarter of a tank. Panic gripped him and he. . . .

A hypothetical scenario. Well, dress rehearsals for it already have been held in scores of communities. A full-blown shortage of massive proportions could develop if America doesn't prepare for a period of much tighter supply, of much higher prices and of much greater dependence upon foreign suppliers, mainly Arab. In 1973, shortages already were being felt in certain areas. Chicago's O'Hare and New York's Kennedy airports could not supply enough fuel to airlines to meet the demand. Planes had to refuel elsewhere. Fuel shortages in Iowa, Minnesota, and elsewhere forced factories into short working schedules. At various times, filling stations were on short schedules in most of the states, while some stations sold gasoline only to regular customers.

These were only faint warning signals of what may lie ahead. This I pondered as I sat in the anteroom of Sheikh Al-Otaiba's office with other journalists. An Arab in robes, red *keffiyeh* on head, poured cardamom-flavored coffee into tiny cups from a highly polished brass pot. It had a spout like the beak of a toucan. A small tin crescent hung at end of the spout, the mark of the brass workers of Hofuf. Once, I visited this ancient Saudi Arabian town amid the date palms of the largest oasis in the world. There a master craftsman explained that the small ornament on spouts of Hofuf pots made them instantly recognizable in the Arab world—a trademark that went back to ancient times.

I took one of the small cups, enjoyed the warm, highly spiced brew that the coffee man poured for me. A cold wind blew off the Gulf and somewhere a window was open. Though the room wasn't uncomfortable, it could have offered a bit more heat. So I extended my cup again, then again, remembering to shake the cup lightly after I gulped down the third offering. Otherwise, the waiter might have continued to fill my cup all night.

Arab hospitality follows rigid rules that dictate the comport of the guest as well as of the host. Custom calls for the host, or his servant, to continue to offer drink or food until the guest has been satiated. It also calls for the guest to drink or eat what he is offered, even if it must be in token amounts. One denotes a "no more" with a cup by rocking it gently in the hand. This avoids the need for voicing a refusal.

An aide from the Ministry of Information hurried in, robes sweeping the floor. He nodded to a reporter from a Beirut paper and said, "Allah is good. We got everything we wanted."

Even before the press conference started I could guess what he meant. Earlier in Riyadh, Saudi Arabia, I had listened as Sheikh

Ahmed Zaki Yamani, Saudi Arabia's Harvard-educated minister for oil, explained how Arab nations were taking 25 percent of the equities of producing oil companies for compensation. Countries would move to 51 percent control on January 1, 1982. (Many sources aver national control will come much sooner.)

Negotiations had been underway for months between representatives of Arab oil producers and oil companies. The countries involved were Saudi Arabia, Abu Dhabi, Qatar, Kuwait, and Iraq. The company negotiating team was led by George Piercy of Exxon, Albert Decrane of Texaco, and Michael Pocock of Shell.

It was obvious that Sheikh Al-Otaiba meant to announce the results of those months of negotiations. Major oil companies would know the price they would pay in lost revenues. Consumers in America, Europe, Japan, and elsewhere would hear the news through price increases in 1973 and the years thereafter.

It was nearly 6:00 P.M. when we filed into Sheikh Al-Otaiba's elaborate office. A 35-millimeter movie camera sat on its tripod in the center of the room. It took most of the available space not occupied by the divans and chairs found in every Arab office. The Arab executive often has a dozen or so of his aides on hand through most of the working day, each man ready to contribute his bit of knowledge should an occasion arise.

Sheikh Al-Otaiba sat behind a rosewood desk with elaborate gilt-gold scrollwork. Aides stood on each side of him, all wearing the Arab robes of the Persian Gulf, headdresses so neat that they might have been changed just before the conference. An Arab male seldom removes this headdress even when inside a building.

The sheikh rose, reserved, quiet looking, yet forceful. He wore white headdress, a brown robe over a dark blue suit, and an air of command. His fingers played with his goatee as he nodded to reporters. He shook hands with each one of us, weakly squeezing mine. The firm handclasp is considered uncouth among many Arabs, the mark of a man who has more muscle than brain.

"Eh, New York!" he said to me. "You have come a long way. We are honored to have you with us."

Small talk. Yet the cultured Arab injects sincerity into such a remark. Hospitality is part of the Arab character, a relic of the hard desert life where you fed a stranger one day, for you might be the stranger the next.

The Sheikh sat down, and so did everybody else. Without preliminaries, he seized a microphone.

"Participation has become a fact," he began, staring into the humming movie camera. "An agreement was reached after midnight this morning in Riyadh under which we will assume twenty-five percent participation in the oil companies producing within our country."

There was an audible sigh. This was it. Western oil companies had dominated the Middle East since the first discovery well was drilled at Masjid-i-Salaman in Iran in 1908. Now their reign was ending.

Often when one sees history in the making there is no realization of momentous events being underway. One needs the glow of time to illuminate yesterday's happenings. But here I didn't need time or that whirring 35-millimeter camera to remind me that an era was over. And what an era. In 1927, oil companies discovered the rich Kirkuk Field in northern Iraq. In the same year, the major companies autocratically encircled a map of most of the Ottoman Empire in their Red Line Agreement and pledged not to act individually "directly or indirectly in the production or manufacture of crude oil" within that area. Then oil-rich Kuwait went into production after ten dry holes in a row; then Saudi Arabia, Abu Dhabi, and Qatar. These were magic names that conjured up pictures of gushing oil wells with Exxon, Shell, Gulf, Texaco, Mobil, BP, and Socal labels on them.

Oil has always been a partnership of sorts, the oil-producing countries serving as junior partners, the major companies as senior partners. Now the junior partners were taking charge. The companies were finding themselves in the position of honorary chairmen of the board.

Participation is the new word in the oil vocabulary of the 1970s. In simple terms, Saudi Arabia, Kuwait, Abu Dhabi, and Qatar were taking a 25 percent equity in the producing oil companies within their borders. For this they were paying a price equivalent to the depreciated book values of the companies, with an allowance for the inflation that has occurred since original investments were made.

That was a bargain for the oil countries. In some cases the price-earnings ratio was only two to one. Oil ministers of countries argue that the original agreements favored companies, with nations greatly underpricing their oil. Participation merely adjusts the situation to realities, they say.

The 25 percent takeover was only the first step. For about two billion dollars, nations by 1982 would take 51 percent of properties that companies say are worth at least four or five times that amount. Still, it is useless to argue merits or demerits of participation. It is a *fait accompli.* The era of company domination is ending and the implications are clear: companies are through as owners of Middle East oil on a concession basis. Already Algeria has nationalized 100 percent of its oil. Iraq seized 100 percent of Iraq Petroleum's operations in the northern part of the country. Iran nationalized its oil as long ago as 1951 but let companies operate for the government so liberally that companies hardly were inconvenienced. Now Iran, too, has taken control of its oil in fact as well as name, while Libya has used nationalization on specific companies as a surgeon might employ a lancet.

Petroleum companies face painful adjustments everywhere in the Middle East, North Africa, Nigeria and elsewhere. Traditionally, the profits of oil were in its production, and subsidiaries that lifted oil from the earth or from sea beds prospered. Downstream merchandising subsidiaries earned only a small profit, or sometimes lost money. Some of this may have been bookkeeping gerrymandering. If a company sells a commodity from one subsidiary to another, it may dictate which of those subsidiaries may take the profit. Usually this has not been the merchandiser of the oil.

With production profits being seized by oil-producing countries, the companies are forced to depend upon downstream operations for their earnings. This means higher prices at the gasoline station and elsewhere. It would not surprise some economists if you paid twice as much for your gasoline and fuel oil in a few years. If inflation continues at the present rate, prices may be higher. The Shah of Iran suggests ten time price hikes are in order.

Price increases may reduce the possibility of actual shortages of fuel. Obviously, there will be less driving when gasoline is $1.50 a gallon than at, say, 50 cents a gallon. Thus, shortages of cash for motoring may develop rather than an actual dearth of fuel.

The inflationary effects of higher oil prices cannot be overstressed. Costs will soar for the mechanized farmer who depends on tractors, combines, and other machinery. Airline fares of the future also will reflect higher fuel costs. Electricity will become more expensive too.

Some may ask, why not just produce more oil in America and forget about those foreigners who are gouging us? Well, there is not enough oil in easily accessible places in America for this country to meet domestic needs. We are consuming it faster than we are discovering it. At the present level of crude-oil production and with no new additions to supply, the United States will be out of oil in ten years, according to American Petroleum Institute figures. Over the next few years, some new oil will be discovered, which will delay the complete exhaustion of American oil. But it will not prevent the need for massive imports.

In the United States, the total energy consumption is rising at a 4.3 percent annual rate, but oil consumption rose by 7 percent in 1972. Meanwhile, our oil production increased by only 1 percent. Already we are importing a substantial volume of oil to make up the difference. That proportion must rise unless Americans reduce their motoring and drastically lower their living standards. Friendly Canada can supply only a part of those imports.

Even with much greater effort and the addition of Alaska's oil, American production is likely to show a steady decline. A few figures might be cited for benefit of chauvinists who won't admit that America's oil heyday may be over. The average U.S. oil well produces only 19 barrels of oil a day, as against five thousand to six thousand barrels a day for the average Middle Eastern well. (A barrel equals 42 American gallons.) Whereas it may cost $2.00 to produce that barrel of oil in America, the Middle Eastern well produces the same amount for 11 to 15 cents.

Economic factors favor Middle East oil when these are given free play. If such freedom is denied, then, America may pay a high price for controlling its market. Such control may be necessary to a certain degree for defense reasons, nearly everyone admits. But the buyer in a sellers' market shouts in a whisper, and America is such a buyer.

Energy problems certainly will hasten the development of competitive energy sources. Nuclear power will be stimulated, and coal may stage a comeback, particularly in areas where pollution problems are secondary to costs. Colorado oil shale and Athabasca oil sands may become sources of some fuel, but not just yet. A science of fuel economy will appear. But technology still will not overcome the world's growing dependence upon Middle East oil much before the 1980s, if then.

"The competing fuels are unlikely to achieve any substantial substitutions for oil in the 1970's," says Sir Eric Drake, Chairman of British Petroleum Company, Ltd., London, a knowledgeable industry spokesman.

The political implications of the energy gap are immense. For over a quarter of a century the Middle East has been mainly an Israeli-Arab arena, to news media. Nothing else mattered as this area wrestled with a problem that seems as insolvable today as it was in 1948, 1956, 1967, or October, 1973. Now the Israeli-Arab question may hold second place for America to energy problems.

The United States has sided consistently with Israel in the Middle East, though the American government claims it is not anti-Arab. This stand is easy to maintain when there are no direct backlashes. But what will be the reaction of the average American if he finds that Arab antipathies prevent him from driving his automobile or from warming his home in winter?

This might prompt him to study why differences exist in the Mideast. He could support possible solutions based upon his own troubles rather than upon somebody else's.

These are only thoughts raised to focus attention upon some of the political implications of the energy gap and of possible Arab responses. An America dependent upon Arab lands for oil, however, may have to reconsider its stand on the Middle East.

It also is self-evident that energy problems will loom larger and larger for the man-in-the-street in America as this decade moves along. At some point, energy may surpass Israel in his mind.

Otto N. Miller, chairman of Standard Oil Company of California, created quite a fuss when he wrote a form letter to the company's 262,000 stockholders and 41,000 employes urging them to show "understanding on our part of the aspirations of the Arab people and more positive support of their efforts toward peace in the Middle East."

You might have thought that "peace" was a dirty word. Somebody tossed red paint on to Socol's headquarters in San Francisco and Jewish organizations protested vigorously. Yet Miller was far from being alone, though it should be emphasized that nobody is calling for an anti-Israel American posture or for abandoning Israel.

"We will need to take a new look at all our foreign policies with respect to the Middle East and attach to them a much higher priority than they have thus far been accorded," says John G. McLean, Chairman and Chief Executive Officer of Continental Oil Company, in New York City.

"We shall have to remember that our domestic economy will be vitally dependent upon peace in that troubled area and upon continuity in the flow of oil supplies. . . . That our friends in Western Europe and Japan will likewise be heavily dependent upon the Middle East for their oil requirements. And . . . that Russia will be the only major world power in the coming decade that will be self-sufficient in energy resources."

That last reminder raises some disturbing security questions, too. America's energy gap is even wider in the natural-gas sphere than it is with oil. Already, at least two giant projects are planned for bringing Soviet natural gas to the United States via liquefied natural-gas maritime tankers. Should we become dependent upon a Communist country for a substantial part of our natural gas? Can we depend upon the Russians to subordinate political considerations for economic? Our energy gap may force us to become more dependent upon Russia, even as food shortages in the Soviet Union may force Russians to become more dependent upon us.

Those who distrust Arabs on general principles worry about our growing dependence upon the Middle East for oil. There have been four wars in this area since 1948, plus countless military scrimmages. American support for Israel has won no friends for America in the Arab world. So Arabs will become our main oil suppliers!

America will be in difficulties if they don't. Thus, we should get to know those Arabs better. We must try to understand them, and what makes them tick. We should try to reduce our differences by first recognizing their causes.

This, of course, does not mean "selling Israel down the river." Israel is a country in the Middle East which is likely to be there just as long as there is an Iraq, a Syria, or an Egypt. But the situation could involve harkening to the plight of the Arab refugees. It also might imply trying to soften some Arab viewpoints concerning Israel.

Thus do oil and politics merge, even though producing nations such as Saudi Arabia have tried to separate them, futilely it now appears. Oil has always mixed with politics, especially in the Middle East. Even as the first discovery well was being drilled in Iran, Russia and Great Britain were maneuvering for diplomatic advantages in this land which was once known as Persia.

Today a Communist rather than a Czarist Russia still maneuvers for a leading role on that stage, proving that the complexions of oil players do not change despite ideological revisions of the casts.

Since Egypt's President Sadat expelled the Russian military from his country in 1972, the U.S.S.R. has lost diplomatic ground in the Mideast. They are trying hard to regain mileage as a result of the Yom Kippur War. They have roots in Syria and in Iraq, and hope to gain a foothold in Libya should something happen to El-Qadhafi, the young revolutionary who opposes Communism as vigorously as he uses oil money to undercut Israel in Black Africa.

The Russians want a finger, then a hand and finally an arm in Middle East oil. They realize that this petroleum is vital to Western Europe, America and Japan. With control of this oil, Russia could dominate the world. One suspects that World War III, if it comes, would be aimed at frustrating any such oil dreams the Soviet Union might have. America just would not let those dreams come true.

Russia probably knows this only too well. This makes it highly unlikely that the Soviet Union would risk any war by overtly intruding in the Middle East. Subversion is another matter. The sheikhdoms and kingdoms of oil are ripe targets for the various movements the Soviet Union supports to foster its aims. America learned in Vietnam how difficult it is to meet subversion with military might.

The Soviet Union itself is self-sufficient today in oil, if not always in grain. It may need some outside oil in a few more years, however, to help its Eastern European satellites meet their energy requirements. Meanwhile, higher oil prices favor the Soviet Union. Thus, it may be expected to support any Arab move which pushes prices upward. It may be significant that in those negotiations for selling natural gas to America, Russians quote prices three times higher than 1972 American prices.

The effects of the energy gap could be catastrophic on the United States balance of payments. This is the difference between all earnings and spending in foreign markets. The United States had a perennial balance-of-payments deficit through most of the 1960s, and into the 1970s. This is why the dollar was devalued twice in fifteen months—in December 1971 and in February 1973. As America's consumption rises (energy consumption by 4.3 percent, oil consumption by 7 percent in 1972), the balance of payments is being hurt. Rising consumption means more imports. More imports mean more spending, even at fixed prices. But prices are rising, and quickly.

In 1972, the United States fuels trade deficit (the dollar difference between exports and imports of fuels) amounted to an outgo of about three billion dollars. By 1980, if current trends continue, the energy deficit may be in the range of fifteen to twenty-one billion dollars annually. This could grow to an annual level of thirty billion dollars in the 1980s.

Traditionally, when nations encounter balance-of-payments problems, the currencies of those nations deteriorate in value. It is evident that fresh strains will be put on the U.S. dollar unless America can develop some means of paying for the oil and natural-gas purchases. America's total exports of goods and services in 1972 amounted to about fifty-seven billion dollars, indicating the scope of the problem lying ahead.

America, Western Europe, and Japan will be competing energetically to get as much Arab oil at as reasonable a price as possible. They will be competing even harder in the world export field to earn the foreign exchange needed for that oil.

And what about the Arab nations? Like bankers, they will be collecting money from all directions, becoming financial powers in their own right.

The next decade and a half may be another Golden Age for much of the Middle East and its peoples. The Middle Easterner, of course, is an abstraction, for he has no concrete unified identity. He is many-faceted, sharing Islamism or Arab culture, or both, with his neighbors, except in Israel. Yet Israelis are Middle Easterners, too.

Nature and man have dealt harshly with this region between the eastern Mediterranean and Afghanistan. Yet, nature did bestow the oil on it in a way that promises to bring it economic salvation, except for Israel. What the oil means to Israel is not yet clear, though if real peace could be found, there is no reason why Jews couldn't be prosperous, too.

It may be an accident of geography and of history which has mated the Middle Easterner and oil at this particular point in man's recorded history. But the Middle Easterner has had an affinity for linking himself with dramatic turns in the history of man. In fact, man's recorded history began with the Middle Easterner.

II

Oil and Land in the Making

I spread the map of the Persian Gulf on the table in the lobby of the Gulf Hotel in Bahrain and studied it with Ahmed, the lean, robed official greeter who had been dispatched by the Ministry of Information to offer assistance. He was an English-educated Arab of about thirty-five, as shrewd as a black jack dealer at a thriving casino, and equally skeptical.

He had told me a little about his island home, how it was the first state on the Gulf to produce oil, and how the revenue provided educations for the Bahrainis a generation before petroleum benevolence affected other states in the area. Now, Bahrainis sometimes seem to be three generations ahead of everybody else on the Gulf in perception. In Dubai I had found that the ruling sheikh's chief adviser was a Bahraini. In Oman, I encountered Bahrainis firmly entrenched in the ruling bureaucracy. Oil companies from Iraq to Sharjah have them on staffs in positions where brains rather than brawn are in demand. Bahrainis have an appreciation for intelligence and overt scorn for stupidity.

"I suppose," I said, "you might term the Bahrainis 'the intelligent elite' of the Persian Gulf?"

A pained expression crossed his angular features. I could see immediately that I had said the wrong thing.

"It is not the Persian Gulf," he said, with a precise selection of syllables. "It is the Arabian Gulf."

Red faced, I nodded. I had known that the Gulf is called by one name on its Iranian side, by another on the Arab side, but I had not realized how prickly the topic might be to a proud Arab. In the Western world, it is called the Persian Gulf and so it will be in this book, with profuse apologies to Ahmed and every other Arab who might be offended. Oil men surmount the problem by referring to this body of water as The Gulf, an inoffensive term that upsets no one, and is recognizable to everybody.

Even the geographic name for the Middle East is not consistent. The term "Middle East" is a late-blooming title for a region that once was known as the Near East to Europeans, who thought the whole world revolved about their continent. They separated Asia into the Near, Middle, and Far East, meaning "near," "middling," and "far" from Europe. The term Middle East as used today has no more connection with "middle" anything than New York has with anything new. Although the term Near East is still used occasionally —especially by scholars and intellectuals who want to make sure that you note their credentials—one seldom hears it in the Middle East any more, especially in petroleum circles.

Generally, the term Middle East designates that part of the globe lying between the eastern shore of the Mediterranean, and Pakistan and Afghanistan. Generally, Egypt, a North African country, is included. Libya more properly belongs in the North African orbit geographically. However, because it is an Arab oil-producing nation, it often is included in statistical compilations of oil production, albeit separately—i.e., "Middle East and Libya."

It is an area that includes twenty-three countries and sheikhdoms, if Cyprus is considered as a part of it. Moreover, to add to the difficulties of neat geographic tabulation, seven sheikhdoms are included in the United Arab Emirates, that cluster of little states on the Persian Gulf that used to be known as the Trucial States. Despite the federation of these states, each maintains its own autonomy. And each is different. Arab unity still is only a wistful dream, no nearer reality than it was in the 1860s, when the Syrian Scientific Society sparked a rebirth of pride in Arab history, literature, and culture. Arab nationalism flows as many streams, rather than as a strong river.

In fact, Arabs sometimes seem to spend more time fighting each other than in finding avenues for cooperation. In September 1962,

Egypt, for example, intervened in a civil war in Yemen. For five and a quarter years its troops fought and died there in a bloody war in which Arabs killed Arabs. From August 1971 into 1973 Syria had no diplomatic relations with Jordan. Moreover, Syria barred its highways to Jordanian-bound traffic. In 1961, Iraq seemed intent on absorbing Kuwait despite the latter's objections. Only British troops prevented the takeover. Oman and South Yemen have been in a virtual state of war as South Yemen supports rebels in Oman.

There are cultural, social, and racial differences. Assuredly, people all belong to the tribe of man. Discrimination because of color is almost nonexistent. The culture and heritage of individuals reflect the many factors that have influenced this region since man first appeared.

Ali Atiyah, a lanky, bearded patriarch in flowing robes at Tabuk, Saudi Arabia, is an Old Testament character with sun-wrinkled features and the wary aspect of a man who realizes that the mere art of living can be a dangerous adventure. He still lives in a tribal tent among others of his kind. A brass coffee pot with long spout bubbles permanently on a small Coleman stove in the tent, ready for the wandering stranger. He is a Saudi Arab.

In Beirut, in an executive office of a bank, Nadia A. El-Khourey, a hazel-eyed, decisive, blonde mother of five, leans back in her swivel chair like the chairman she is and briefly sketches sales trends of the many companies in her business empire. She is an anomaly in the Arab world, where a woman's place not only is in the home but in a quiet corner of it. She is a Lebanese.

In Tel Aviv, Eliav Simon, heavy-set, in his fifties, with thinning gray hair and the courtly manner of a Spanish grandee, leans across a table in the Chez Christian Restaurant and explains how modern Tel Aviv is overrunning the old city of Jaffa as builders push ever outward. He speaks Hebrew, English, Arabic, and French fluently, holds a degree from American University of Beirut, and can trace his Sephardic family line through many generations of living in Jerusalem. He is an Israeli Sabra.

In Tehran, Dr. Manoutchehr Eghbal rises from behind his desk in the headquarters of the National Iranian Oil Company and extends a hand. He is crisp, alert, of imposing physique, with graying, curly hair, who looks somewhat like pictures of Red Grange, the

American football player of an earlier era. He is a French-educated medical doctor who has filled just about every government post— prime minister, minister of the interior, minister of communications, and minister of health. Now he heads NIOC and is his country's chief oil man, a suave, many-talented man who is equally at home discussing the plays of Molière, the economics of Persian Gulf oil, or the latest methods of treating the El Tor strain of cholera. He is an Iranian.

In Kirkuk, Iraq, Ibrahim Qadi Barzani is a long-moustached merchant of plump build who looks even fatter because of the baggy Turkish pants he is wearing. Sitting on a stool before his open-fronted shop, he puffs contently on a water pipe, red tarboosh on head, as he reports that two of his sons are attending the university at Baghdad. He hopes, though, that they will not come back with dissolute habits acquired from the Arabs in the capital. He is a Kurd.

The Middle East has Copts, Druses, Nubians, and Armenians, in addition to Arabs. Among Arabs there are Maronite Catholics as well as Moslems. Expatriate Pakistanis, Indians, Americans, and Europeans add to the Babel of peoples.

An alphabetical listing of the countries and sheikhdoms of the Middle East has a romantic ring to the ear of the world traveler— actual or armchair—as it proceeds from Abu Dhabi to Yemen. It is a land of wide deserts, where the sands may blow as hard as dry snow in a blizzard. Mountains rise stark from arid plains. Barren rock changes colors with the sun from a harsh yellow to pale pink, then to the deep purple that heralds a star-filled night. The Red Sea, the Arabian Sea, and the Persian Gulf bear dhows with lateen sails, with turbaned crews from pages of the *Arabian Nights*.

Everywhere in the Middle East, long-dead cities of marble, or more likely, of clay brick, invite the archaeologist and the curious traveler. There are old, still-inhabited cities where the rubble of millennia lies beneath modern pavements.

There are also holy cities such as Jerusalem for Christians, Arabs and Jews. All three faiths revere the old city imprisoned behind its limestone walls, dominated by the golden Dome of the Rock Mosque.

At Sidon in Lebanon, we followed a guide through narrow lanes of the old city. An arched passage tunneled through a jumble of buildings. Women shopped at an open-air fruit market. A gap in buildings revealed a view of the old Crusader castle in the harbor.

"This was the land of Canaan of the Bible," said our guide. "Anywhere you dig you find an old building, a very old building."

"But no oil?" I said.

"No. No oil in Lebanon," he said.

And there are glittering, modern cities with high-rise offices and luxurious apartments, such as Beirut, Lebanon, set magnificently between sea and mountains. "This is the capital of the whole Middle East," Samyr Souki, a Christian Arab business consultant with the sophistication of a Madison Avenue executive, said. In his modern office a secretary waits at his elbow for dictation, a telex connects him with the outside world. Casually he talks about a forthcoming business trip to New York as if it were a frequent occurrence.

Such are the diverse pictures that come to mind when that alphabetical listing of the Middle Eastern states is cited: Abu Dhabi, Ajman, Dubai, Bahrain, Cyprus, Egypt, Fujairah, Iran, Iraq, Israel, Jordan, Kuwait, Lebanon, Oman, Qatar, Saudi Arabia, Sharjah, South Yemen, Syria, Ras al-Khaimam, Turkey, Umm al-Qaiwain, and Yemen.

Startling monuments and old cities from antiquity are found in every country of the Middle East. Few things are more awesome than to be riding on a bus when it swings around a naked, sandstone ridge on the shore of Lake Nasser in southern Egypt, to find yourself facing Abu Simbel. Three gigantic seated figures stare outward from the wall of a cliff which has been transformed into a temple in the solid rock. The head and torso of a fourth figure has fallen. That doesn't detract at all from the immensity of this monument built by the Pharaoh Ramses II about 3,200 years ago.

"There is nothing like this anywhere else," said Hosni Ali Ghoneim, a 34-year-old Cairene who has made a deep study of the archaeological relics of his country. He spoke with that pride which one notes at Persepolis when an Iranian talks of the monumental staircase to the palace built by Darius the Great, or which the Iraqi voices at Babylon before the Ishtar Gate. The Middle Easterner views those monuments as his own, even though a hundred or more generations may have so diluted blood strains that modern peoples may have little relationship to the ancient builders.

Turkey, a country which dominated the Middle East for centuries, today, has little affinity with the Arab lands which were its satraps for so long. It plays no part in the current oil drama.

Fourteen of these states produce some oil, Israel from occupied Sinai. All of the major oil exporters touch the Persian Gulf. (Libya, which is geographically part of North Africa, of course, does not.) Oil and the Middle East seem synonymous. Actually, petroleum of this region is highly concentrated in the Gulf.

In the School of Sciences at the University of Riyadh, exhibits and a relief map of the Arabian Peninsula relate the geological story of oil. We sat in a class of twenty oil company trainees who were participating in a three-day seminar. Lectures were in English, the lingua franca of oil in the Middle East.

"Allah was generous when he gave us the oil," a geology professor who did not look any older than his students, began his lecture, as students listened raptly.

One hundred and fifty million years ago the oil-bearing area of the Persian Gulf was a tropical land of forest and swamps. Lush vegetation covered ground that must have been watered by equatorial-like rains. Giant ferns formed thickets around bases of tall trees. Huge reptiles and strange creatures inhabited the grassy marshlands, creatures with names such as the ichthyosaurus, the dinosaur, the marine ganoid, and the stegosaurus.

The professor spoke slowly, probably realizing that, though these students understood English, they were not masters of the language. Students listened intently. The art of story telling is appreciated in the Middle East. Even today villages pay homage to the itinerant story teller who may regale youths and elders with tales of Harun el-Rashid, of the Arabian Nights, or of the Conquests. Oil becomes a dramatic topic in an Arab classroom.

He went on to relate how broad rivers swept down to the seas of that time. They carried the sludge left by the decomposition of the rich vegetation. These were shallow seas that teemed with marine life, from tiny plankton to huge marine monsters. Seaweed formed underwater forests for these water creatures. In them, countless generations of marine life lived, died, then sank to the muddy bottom to be covered by the sediment of rivers. That sediment grew deeper and deeper. The weight of the mass above helped create sandstone below. Salts in the sea transformed some of that sandstone into limestone and other minerals.

Bodies of the long-dead sea organisms and the vegetation slowly decomposed. Pressure and chemical change transformed the carbonaceous masses into what we now know as petroleum.

"The whole process probably took about sixty million years, from perhaps about one hundred and thirty-five million before Hegira to around seventy-five million before Hegira," the lecturer said.

With such figures, the reference to the Moslem calendar instead of the Gregorian meant little. It still was 135 million B.C. and 75 million B.C. Reconciling science and Allah sometimes is a problem in Arab lands. The fanatically orthodox Moslem does not allow much room for scientific teachings, which seem to conflict with Islam. The modern Arab reconciles his beliefs with Islam just as do Christians with the Bible and science. He feels that the basic tenets of Islam do not conflict with science. It is the interpretations of little men that cause big trouble.

These nonuniversity students had the equivalent of an American eighth-grade education. Most worked in the oil fields. Now they were being force-fed university data in a cram course.

"Why did Allah take so long to make the oil?" one student wanted to know.

The professor deftly fielded the question. He was a Moslem himself and undoubtedly he had adapted his thinking to the modern age without sacrificing basic beliefs. Even if he had been a nonbeliever he probably would not teach long in an Arab university if he were to betray his feelings.

"What is sixty million years to Allah? It is but a drop in the bucket of time," said the professor.

"Were there men in Arabia at that time?" asked another student.

The professor shook his head. "Allah had not yet created man as we know him today."

He explained that man came upon the scene long after the geological processes occurred in the earth's laboratories. Geological changes still are underway, for the earth is a dynamic ball that has never known a settled period, and never will until the planet is dead.

After the classroom session the professor introduced himself as an Iraqi expatriate on contract in Saudi Arabia. "You must visit Iraq," he said. "It is the cradle of mankind."

We already had visited Iraq a dozen times over the years. It is the place to follow the trail of the Middle Easterner as he first appeared on the stage of history.

In Iraq, artifacts and sites of ancient history are never far. This was Mesopotamia, the land along the Tigris and the Euphrates rivers. Here, legend says, the Garden of Eden once flourished. Here nomadic man settled into an agricultural existence that produced the Sumerian, the Akkadian, the Amorite, the Chaldean, and the Assyrian civilizations. Here was ancient Babylonia, where, the Bible tells us, the Jews were held in captivity.

For generations of Europeans the Middle East meant archaeological digs. Scholars paid more attention to the rubble of ancient civilizations than to modern Middle Easterners. It took oil to add another dimension to the picture.

Do these old civilizations have any connection with the modern Middle Easterner? The answer is yes. We are all part of the same stream that flows sometimes turbulently into those obscured canyons of the future. The Middle Easterner lives a little closer than most to sites of those original springs from which gushed man's culture. Those springs nourished three of the world's great religions; the arts of writing, sculpture, painting, and architecture; the sciences of agriculture, geometry, and astronomy.

The Egyptian *fellahin* still till their land according to cycles and practices developed in the time of the pharaohs. The mud brick houses of Iraq had their counterparts in the ancient cities of Mesopotamia. Designs of the Beau Geste forts that are so common in Arab lands have their roots in the walled cities of the two rivers, a factor evident in any study of the Ishtar Gate of Babylon.

Educated Iraqis, Egyptians, Lebanese, and others understand their relationship to the past. To the illiterate peoples, the connection is vaguer, obscured by the weight of the Islamic culture, which has been superimposed atop the older structures. Any time prior to Mohammed is the Time of Ignorance, a period when the word of God was being heard only by a scattering of prophets. Yet the illiterate, too, are bound to their ancestors either through their genetic strains or through customs and practices which have roots in the historical dawn.

Today an American comes to Iraq with some visa difficulties which were unknown in those early ages. Iraq severed diplomatic connections with the United States during the Six Days War of 1967 as a gesture of solidarity with the Arab cause. America now is identified as a military ally of Israel, and to an Arab the "friend of my enemy is my enemy."

"Don't worry. You will have no trouble," an Arab friend on *Al Thawra*, the country's biggest daily newspaper, told me in Baghdad. He was right. The rhetoric of the Arab seldom explodes into anti-American violence against persons, though the murder of American diplomats in the Sudan proved a bloody exception.

But even as my friend talked I knew that an Arab usually is supremely optimistic everytime he starts a trip. Part of this is due to the fatalism of the Arab. What will be will be. Man shouldn't worry too much about what is in Allah's hands.

In Beirut, once when there were demonstrations and shooting in the streets, I asked a cab driver at the Phoenicia Hotel if it was all right to drive to the Post office, to dispatch a cable.

"No trouble," said he, shaking his head.

So we drove through near deserted streets, past shops with shuttered fronts. Armed guards searched incomers at the Post office. A sandbag redoubt protected the front door. The clerk accepted the cable and I returned to the taxi.

Then shooting started up the street. The driver made a quick U-turn, stepped on the accelerator. Tires bounced over the rutted pavement. We hurtled toward the waterfront and the promenade drive on which the Hotel Phoenicia sits.

"I thought you said there wasn't going to be any trouble," I said.

"For us there was no trouble," he said, cheerfully. "We have returned. If you go expecting trouble, then you may find it. Better not to think about it."

In Iraq one thinks of past civilizations. Mesopotamia, the land between the rivers, became the motherland for a series of civilizations which left their mark upon every subsequent era. This was the

Fertile Crescent, a quarter-moon-shaped area with one tip pointed south at the head of the Persian Gulf, the other pointed south at the Mediterranean.

"The term 'Fertile Crescent' was invented by James Henry Breasted, the American historian," my journalist friend at *Al Thawra* said as we sipped mint tea on a terrace of the Baghdad Hotel. There was bitterness in his voice as he added, "Outsiders have always tried to shape our lives and our land for us."

"So now you are a Baathist," I said. He was an articulate voice for this left-wing political party, which preaches that Arabs should unify and rule themselves. Their oil is a personal possession with its ownership a matter of honor.

"Yes," he said, and as if he had read my thoughts, added, "We no longer can allow outsiders to dominate our oil."

I had first known him as a correspondent in Beirut years before for a news wire service. Like many Iraqis, he was hot-tempered, quick to see offense where none was intended, generous with both his time and money, and a man of his word. When the check arrived, he would not let me take it.

"You are my guest," he said. He glanced at his watch, grimaced, then cheerfully decided to forget time. "I am late for my appointment now, so I might as well forget it and help you line up an automobile to visit some of our historic cities."

"You make me feel guilty," I said.

"Don't," he said. "No one keeps an appointment anyway. One hour. Two hours. Does it make much difference?" Then his voice changed. "I must bargain for the car. Otherwise you, as an American, will be paying five times the going rate."

"Are they that gyppy?" I asked.

He shook his head. "It is not gypping to get as much as you can in a deal. If the buyer is stupid, it is not for the seller to cry."

He lined up the automobile, with charges specified in writing, obtained a letter from the Archaeological Department introducing us to various directors at sites, and marked several trips on a map.

His parting words were, "My country is the homeland of all civilization, and I want you to appreciate it as I do." He handed me some literature about the ancient civilizations.

Usually Arab publicity material is so bombastic that the propaganda defeats itself, especially in the left-wing states. Hyperbole and overstressing are common flaws. No superlative is ever used if a super-superlative can be found to replace it. Politics may be interwoven into everything from a cook book to Coca-Cola. Often truth will be stretched to cover any holes in the Arab garment. When Israeli commandos kill three Al Fatah guerrillas in Beirut, the shooting has hardly stopped before Arab propagandists claim American CIA agents are involved. Somebody must be blamed. To blame Israel alone is to make them appear as supermen, and Arabs cannot accept that. Nor can they admit that Arab military slovenliness may sometimes be the best ally Israel has.

Then I noted, though my friend had not said so, that he was the author of the historical matter he had given me. It was a simple, straightforward tale that showed a pride in the past accomplishments of long dead civilizations, as in this paragraph:

> Jarmo, in northern Iraq, is the earliest known example of settled, agricultural and non-nomadic life. The wheel was invented in what is now Iraq as was possibly the earliest writing. The earliest known code of laws has been found here; the earliest known astronomical observations were made here; the earliest known university has been found here.

That is quite a collection of firsts, I reflected, when starting the next day. The road north goes to old Nineveh and the Assyrian land where Kirkuk fields produce oil. The road south proceeds to the land of the Sumerians, those people who appeared on the chalkboard of history about 3300 B.C., and to the oil fields of Basra. We took the south road.

The asphalt highway passes through rich, flat fields that must have been fertile in Sumerian times. Nobody knows whence came these short, stocky, non-Semites who wore skirts of flounced cotton. Their language is a speech apart.

"No other language has been found anywhere else in the world that corresponds to Sumerian," one linguistic expert at the Iraqi Museum later told me.

In ancient times, the Fertile Crescent must have seemed like a green garden to nomads coming from neighboring deserts and arid mountains. No wonder that legends placed the Garden of Eden in that area between the two rivers where even today date palms tower over green fields. Civilization flourishes where man has a full stomach and political stability. It dies or stagnates in revolution and violence. This has been the history of mankind since there was a history.

Our driver had the typical Middle Easterner motorist's contempt for everyone else and everything else on the highway. His name was Ahmed. (Every second Arab in the Middle East is named Ahmed, Ali, or Mohammed.) He was a cheerful Shiite Moslem with a family of six children by two battling wives. When I asked him which was his favorite wife, he responded, "Neither." He assured me that women are the snares of Satan. Outside of children, no good could come from them, though a man must accept this, for it is woman's unchangeable nature. Allah meant this to be, so man must accept Allah's will.

As he drove, he talked with his hands, which made driving difficult. Buses and trucks clogged the asphalt road south from Baghdad, each driver hogging the road as if he had a deed entitling him to the highway's center. Arabs are a courteous people, but not when behind the wheel of an automobile. Each encounter with an approaching vehicle was a deadly game of "chicken" with each driver seeking to force the other to turn aside.

Iraqi peasants harvested grain in the fields, long robes dusty with wheat chaff. Minsk-made tractors provided power for Moscow-manufactured grain wagons. Beyond the tawny fields of grain, flat fields of alfalfa lay green in the hot sun.

Clumps of trees marked distant villages. Houses rose in a tight cluster so as not to intrude too much upon the rich agricultural land. Our driver paid no heed to anybody but me. Like many Middle Easterners, he had an instinct for baksheesh. He was more concerned with riveting my attention to assure that I knew he existed, than in making certain that we reached our destination safely.

His favorite expression was the fatalistic *Maktub* (It is written). If Allah willed, we would reach our destination. If Allah did not, then we must accept Allah's will.

Twenty-eight miles from Baghdad, the road entered the sprawling village of Al Mahmudiyah. High compound walls sheltered mud-brick houses with flat roofs, on which families would be sleeping in summer heat. Date palms within compounds provided bits of greenery against the tawny background of sun-dried brick.

Several Iraqi farmers in robes, *keffiyehs* on heads, leaned against pickup trucks parked at angles on the wide street. A crowd shopped at open-fronted stalls. On display were all manner of things—tinware, pistachio nuts, meat hanging on hooks in the sun, and gaudily colored textiles.

"You know Hammurabi?" Ahmed asked as he raced into town.

Of course, I didn't know Hammurabi, personally. He lived 3,750 years ago. But nearly every school child has heard of his code of laws. In Baghdad I had been told that any study of Arab law should start with Hammurabi's Code.

"Sure. I know Hammurabi," I said.

"His town." Ahmed pointed to the floor of the Ford Galaxie sedan as if the Babylonian king still ruled the land beneath the floorboards. Like most of the information Ahmed gave us that day, this was wrong. He had an admiration for the broad elements of his country's history, but no capacity for detail.

Al Mahmudiyah is close to Sippar, the ancient Babylonian empire city of law giver Hammurabi, but several miles separated them. It was in Sippar that the Amorite king erected a basaltic column with his legal code inscribed in cuneiform upon it.

Currently, a few piles of gray mudbrick mark the site of what was once a great city in the early morning of man's recorded history. Ahmed roared through the modern village, horn blowing constantly. He scattered pedestrians, a wayward donkey, and my equanimity.

"Take it easy," I said, almost involuntarily. He missed a pushcart fruit peddler by inches.

"Maktub!" said Ahmed. He nonchalantly shifted his gaze toward me, unabashedly said, "Me good driver. Always people me big tip."

I could not see why he was unable to apply his fatalistic logic to his future tip. Whatever I gave him would be what Allah willed. So be it. His inconsistency was characteristic of the Middle East. A man may show his contempt for his wife in certain ways, then be surprisingly considerate in a situation where he realizes she is the mother of his children. Girl babies are rated so low that friends greet the new father by saying, "I bid you tears." Yet most fathers show love to their daughters, though perhaps not as much as to their sons.

Once in Syria, a businessman from Homs gave me a lift to Palmyra, the oasis city of Queen Zenobia in Roman times. This was during a period of unrest when street demonstrations were occurring almost daily in Damascus, and when left-wing Baath Party zealots were seeking to eliminate all opposition. The businessman, as a supporter of free enterprise, knew that he probably was on the Baath blacklist. Yet he talked philosophically about his position as if he had resigned himself to the will of Allah. He intended to reason with any Baath radicals who appeared, and trust Allah to help him.

Then, he patted the bulge of a shoulder holster under his jacket, and said; "By all means make friends with the dog, but do not lay aside the stick."

There was another time in Beirut during the period of the 1958 American occupation of Lebanon's Beirut airport area. Most journalists filled in idle hours by going into the rebel Arab quarters held by forces of Saeb Salam, the politician who wanted Lebanon to align its policies closer to those of Arab nationalism. Such trips usually always produced stories. Sometimes there was trouble, too, when fire fights developed between rebels and government forces.

On one such trip, I was accompanied by a journalist from a local Beirut newspaper, a cocky young Arab with an apparent contempt for danger. He condescendingly viewed my interest in escape routes in that maze of streets as a definite sign of weakness.

"Trust Allah and you will be at peace," he said.

So once again the shooting started somewhere just ahead. I turned at the first shot, raced back the way I came. Or perhaps I should say that I shuffled along faster than usual. I am not built for racing on any track, though it is surprising how fast a man can move when stimulated by rifle fire. My equanimity was not restored by having my Arab friend pass me as if he wore roller skates.

Later, I saw him on the terrace of the St. Georges Hotel by the sea, practicing the more leisurely bar-room gossip style of journalism followed by so many of our colleagues.

"I didn't notice you placing much trust in Allah when the shooting started," I said.

The remark did not bother him in the least. "But I did," he protested. "Allah told me to run, so I ran."

In Iraq, Ahmed, with his inconsistencies, was not unusual. Over the next few days we covered much historic ground. Babylon, Ur, Erech, Lagash, and Nippur rolled by. These were some of the city states founded by the Sumerians three-thousand years before Christ. Imagination must fill in most of the blanks at sites, for there isn't much to see except crumbled brick.

Sumerians, however, left so much of themselves in their cuneiform writings that we do know a lot about them. Irrigation, the use of money, law codes, all were developed under their aegis. They used wheeled vehicles, employed the arch in building, developed astronomy as a science.

The Sumerians had a keen sense of private property and were ardent free enterprisers in trade. The society was a form of theocratic socialism in that people believed the gods owned the city and the land. Thus, people worked for the particular god who was looking after their particular community, tilling land in communal style. Temple priests, of course, benefited from this system, and probably provided the leadership and the administrative ability to manage for the good of all.

It was not the Sumerians, but successor Akkadians who started Babylon about 2400 B.C. But Akkadians built their civilization upon the base left by the Sumerians, following some of the same religious practices, too.

A replica of the Ishtar Gate marks the entrance to the piles of dried brick, the reconstructed temples, and the uncovered ruins of old Babylon. A caretaker who spoke no English met us before the small museum. He fumbled with an enormous key ring before finding the right one to let us in. That particular day the only other visitor was a French archaeologist and his wife, who spent hours photographing each foot of the Processional Way along which sacred processions used to march during religious festivals.

In the reconstructed E-Mah Temple I could visualize how the even older temples of the Sumerians must have impressed the populace in that distant day. At Ur, even the modern visitor can be awed at realization that, when a king died, his entire court might be put to death and buried with him. To see the glory of the Royal Tombs of Ur, however, one must go to the Iraqi National Museum in

Baghdad, not to Ur. The gold jewelry, diadems, and artifacts illustrate how advanced was this first civilization in man's recorded history.

The theocratic socialism of this society had no relationship to the politically orientated communal states of today. It was not a socialism designed to spread the wealth among many. The Sumerian valued his individual possessions so much that it became common to mark them with individual "brands." A design would be chiseled and ground onto a small cylinder seal, with each man having his own design. The design might be that of an antelope amid vegetation, as was the case with one seal dating from 3200 B.C. It might be a monkey and a dwarf, the goddess Innana with a long-eared sheep, of a bull-man and a bull-warrior, of a worshipper before a god, or of an ibex and eagle below a collection of stars. All of these designs are on cylinder seals dating from Sumerian times.

The cylinder produces the image on it when rolled across a piece of flat, soft clay. When the clay hardens, it could be tied to any particular possession to be marked. Or the clay could be allowed to harden on that property—the front door of a man's house, for instance.

It was a step forward from these designs to the cuneiform writing on clay tablets, probably man's greatest invention. For now, the wisdom of the ages could be inscribed on tablets so that each generation could learn from preceding ones.

Writing fostered development of mathematics, linguistics, literature, and various sciences. Thanks to their cuneiform writings on clay tablets, the record of Sumerian accomplishments has come down to us, providing us with an understanding of how civilization blossomed in the Middle East.

"The art of writing is the mother of orators and the father of artists," says a Sumerian proverb that was recorded on one of those clay tablets. It also is the father of science and the grandparents of culture. Sumerians developed the sexagesimal method of numbering, a system still used today in the division of the hour into sixty minutes, the minute into sixty seconds, and the circle into three hundred and sixty degrees.

Sumerian literature influenced all adjacent and most future civilizations, and probably provided some material for the Old Testament Bible.

Hammurabi appeared on the scene long after Sumerians had receded into historical distance. There is little to see at Sippar itself, the city associated with his code of laws. But the story is well illustrated in the Baghdad Museum. The loving care of national monuments and the professional manner of museum exhibits clearly show that Iraqis are proud of their culture. I had an official letter with me when I visited the museum, and thus obtained royal treatment. However, so enthusiastic was the curator who became my guide that I am sure he would have been equally helpful with anyone who had displayed interest in the stories of the Sumerians, Akkadians, Chaldeans, Babylonians, Assyrians, and other early peoples of Mesopotamia.

"Except for the Sumerians, all of these early peoples were Semitic," my guide explained. He wore a Western suit, which needed pressing. With his thick spectacles and lined features he looked like an absent-minded professor who is more concerned about the intellectual task in hand than about inconsequential matters such as the appearance of a suit. There was pride in his voice when he uttered the word Semitic. He was emphasizing that his ancestors played a part in the creation of these early civilizations.

Semites entered the Fertile Crescent as nomads, probably from Arabia, settled, absorbed the civilization of the Sumerians, then transformed it into something better. Arabs and Jews, of course, are both Semitic, probably of the same stock, if one goes back far enough. The mythology of the Arab people says that the Semites descended from Shem, the oldest son of Noah. The term *arab* means "desert dweller." Since all Semite tribes originally lived in the desert, Jews may originally have been Arab.

Hammurabi, one of the greatest of the early Semite kings, is of interest to anyone seeking to know the Arab. This Amorite king reached backward in time to earlier codes and to the laws of the desert Bedouins to form his law code. (A Bedouin is merely an Arab who follows the old nomadic life in the desert. When he settles somewhere he becomes a Saudi, an Iraqi, or whatever.)

Hammurabi ruled from 1792–1750 B.C., by which time law codes of the area already were well advanced. In the Iraqi museum I was shown earlier codes in cuneiform that preceded Hammurabi by a couple of hundred years. But Hammurabi's code survived on that black basaltic column which now is in the Louvre in Paris. In history it is not always the best or the first that is written in indelible ink. What counts is what survives to later man. Sometimes benchmarks are found long after by the chance cuts of an archaeologist's pick.

"We have all 282 of Hammurabi's laws translated in several languages including English," said my museum guide. "Of course, you know what the Koran says of the thief. That could only have come from the code of Hammurabi."

I had read the Koran. But he had to direct me to Sura V, verse 38: "As for the thief, both male and female, cut off their hands. It is the reward of their own deeds, an extremely exemplary punishment from Allah. Allah is mighty, wise."

I copied down Hammurabi's laws as they were translated directly from the old cuneiform script:

If a man has accused a man and has charged him with manslaughter and then has not proved it against the accused, then the accuser shall be put to death.

If a man is robbed without recovery of his property, then he shall be reimbursed by the state, after the victim swears before a god and the local mayor of his loss.

If a judge reverses his verdict after entering a judgment, he shall pay 12 times the claim to the claimant, and the judge shall be disbarred.

If a man has put out the eye of a freeman, they shall put out his eye. If he breaks the bone of a freeman, they shall break his bone.

Hammurabi's code even went so far as to prescribe the fees that a surgeon could charge. For a successful operation on a freeman, the surgeon could charge ten shekels of silver. For the same operation on a serf, the price was five shekels. On a slave, it was two shekels. But if the surgeon operated on a freeman and the latter died, then the surgeon, if found guilty of causing that death, would have his hand cut off.

Even today the eye for an eye philosophy prevails in the codes of the Middle East. In Iran, drug peddlers are executed. The thief in Saudi Arabia still may lose a hand. In Dhahran, a doctor at the Aramco Hospital told us how he has sometimes treated thieves who had hands severed by authorities. Public hangings occur in Iraq. Some countries behead convicted criminals.

After Hammurabi, other empires came and went. Egypt's pharaonic era paralleled that of Mesopotamia's kingdoms. The Babylonians ruled for centuries only to be overthrown by the Assyrians. Hittites, an Indo-European people, became located in Anatolia to the north. Often they warred with kingdoms of the Fertile Crescent, without ever winning control of the area. In Lebanon, the Phoenicians built port cities, helped develop an alphabet, then colonized places as far away as West Africa. Greeks and Romans followed in the Middle East.

None of these early civilizations had an Arab flavor of its own. Each absorbed Arabs and incorporated them into civilizations, as if Arabia were to remain forever a land of emigrants. It was Islam that fired the Arab imagination with dreams of world conquest.

III

Religion for all Seasons

At the Jidda Airport, a dozen Moslem pilgrims descended from the Saudi Arabian Boeing 737 in bright sunshine. All wore the white, seamless garments of *Ihraam*, men with a bared shoulder, women looking like nuns in white, fresh from a convent. Identical dress assured that each person would be identical before God on this most holy pilgrimage to Mecca, the *Hajj*.

"*Laa ilaaha illa llaah*," called the leader of the group, as his feet touched the tarmac.

"There is no god but Allah," his companions repeated after him.

"*Labayka!*" the leader called out.

Once more his companions repeated after him, "I am at your service, oh Lord."

Umbrellas opened to shield bare heads and shoulders from the strong sun. Together the group trekked toward the immigration counter of the airport.

Once again, the annual *Hajj* was underway. Over a million of the world's 500 million Moslems were traveling to Mecca. Most of them were coming through Jidda, the Red Sea port that is only forty miles from Mecca by a four-lane superhighway.

Arriving ships disgorged passengers at the new harbor set on an island in a bay of the limpid Red Sea. Planes disembarked passengers from as far away as Taiwan, Mauritius, and Morocco. Buses rolled in from Jordan, Iraq, Syria, Lebanon and the sheikhdoms of the United Arab Emirates. One rainbow-colored bus carried a group of

Arabs from the Israeli-occupied West Bank of Jordan, a group permitted to depart for the *Hajj* by Israeli authorities.

Few things are more awesome than is the annual *Hajj*, which brings devout Moslems to Mecca and its adjacent sacred places during *Dhu-I-Hijja*, the holy days in the final month of the Moslem year. A frenzy sweeps through the hundreds of thousands of assembling people as the days near. For weeks, people have been arriving, so fearful of travel foul-ups that some come a month ahead of time. Street jams intensify. The highway to Mecca becomes one long line of automobiles, buses, and open trucks with people piled on platforms.

The time is past when camel and donkey trains carried the devout pilgrims those last miles to Mecca. This does not lessen the power of this holy spectacle, which emphasizes the hold Islam has over its adherents. Mohammedanism is an extrovertic religion in its outward manifestations. The devout man thinks nothing of placing a prayer rug, a garment, or a newspaper on the street to pray at the appropriate times while facing Mecca. The speech of the Moslem is interlaced with phrases thanking God for everything, down to the mere fact of being allowed to live. The pilgrimage itself will become a matter of pride for the pilgrim. It will become a part of his official history, like a government decoration or citation for a meritorious task.

Most of the people living in the Middle East are Moslem. Some are strict orthodox. Others are little more than Moslem in name. All regard their religion in a way that sometimes is difficult for the modern Western mind to grasp. Religion in the Middle East is something to be taken seriously, like strong medication to purge the spirit. The Christian may joke about his particular sect, sometimes in a shame-faced manner to hide his true feelings. One seldom hears the Moslem joke about his religion. When Islam comes under attack, all Moslems draw together.

The Koran says: "Fight in the way of Allah against those who fight against you, but begin not hostilities. Lo. Allah loveth not aggressors." In the Yom Kippur War, Arabs claimed their first strike was only a return to Arab lands, not an aggression.

Moslemism is more than just a sabbath religion to the devout person. It is part of his life. He lives by the Koran, quoting it extensively in his daily conversation to stress that it is governing his actions:

> Never will Allah suffer the reward to be lost of those who do right. [Koran, Yusuf 90]

> God is with those who patiently persevere. [Koran, Anfal 46]

Such comments permeate conversations, appear in writings of authors, interlace speeches of politicians and statesmen, and illustrate legal decisions. Saudi Arabia follows principles of Islam in its constitution and in its judicial and governmental system. Libya's fiery leader, Colonel Mu'ammar el-Qadhafi, governs Libya according to the precepts of the Koran.

To a lesser extent other Moslem leaders heed the Koran in making decisions and in interpreting situations. It is as if President John F. Kennedy had openly used Catholic doctrine to define his actions in the White House, or as if President Richard M. Nixon were to turn to Quaker precepts to codify his decision making.

Anyone who wants to understand the Middle Easterner, his psychology and his attitudes, should read the Koran and study the religion that has been built about it. Moslems regard the Koran as the word of God, as revealed to the Prophet Mohammed. Note that Mohammed is only a prophet, not the object of worship.

In the Moslem view, if a person incorporates the Koran into his family affairs, his politics, his business, and other aspects of his daily life, he is merely paying attention to the word of God. If someone asks him about that, his response is apt to be, "Shouldn't everybody?"

"Our religion gives us that something more, that something that binds us all together," said Hisham Nazer, Saudi Arabia's minister of state for planning.

Sheikh Nazer is articulate, trim, with the air of confidence of one who has attained power at an early age, and used it well. When I commented that it was unusual to meet a thirty-nine-year-old minister, a smile formed around his moustache and he said, "I was appointed when I was thirty-one."

I had met him in his olive-green carpeted, mahogany-paneled

office in Riyadh, Saudi Arabia's capital, when I was investigating the scope of his country's development program. Sheikh Nazer, I had been told, is the man who keeps everything moving in a five-year development program, which is seeing more than ten billion dollars poured into schools, hospitals, roads, communications networks, and other paraphernalia of a modern civilization.

After he had outlined dozens of major programs aimed at bettering the lot of the ordinary citizen, I had raised the question, "Won't education and material progress encourage citizens to reach for something else, for more political freedom?"

"We already have something else," he said, quietly. "Our religion. We believe that we have the framework for democracy and for the protection of human and political rights in the writings of the Koran. You must never underestimate the power of the religion which holds us together. All the money we are spending for development in this country would mean nothing if we did not have the people with us, a people infused with the spirit of Islam."

This was not exactly the answer one expects to receive from a minister of planning, but it is the sort of answer one finds frequently in the Moslem world. Politics, economic development, and religion are so interwoven that a new mosque may be included along with a hospital and a school in building plans for modernizing a mudbrick village. Saudi Arabia is completing a new industrial zone in the Eastern Province on the edge of Dammam. The minaret of a government-built mosque rises amid the warehouses, the power plant, and other facilities constructed to assist manufacturers who might want to locate there.

During the holy fasting month of Ramadan entire countries sometimes grind to a halt. Offices are deserted or understaffed. Letters are left unanswered, should any be delivered. Shipments of merchandise are delayed. Tasks are postponed. All attention is focused upon complete abstinence from food and drink from sunrise to sunset during this month. This is a difficult task for anyone who might be doing an ordinary day's work. When the crunch comes, the day's work suffers. Sometimes dispositions may suffer too.

Cairo is a teeming, friendly city crowded beside a Nile that still can look romantic, though high-rise monuments of concrete and

steel now line river banks. Normally, Cairene hospitality is as inviting as the tantalizing cooking odors that float from sidewalk restau-
rants in the Khan el Khalili Bazaar.

But one recent afternoon in the Ramadan a crowd was collecting
on Soliman Pasha Street, that thoroughfare which has some of the
most exclusive shops in the city. Automobiles braked, then honked.
A red bus halted, disgorging passengers who joined a mob pressing
forward. Two blue-coated policemen forced their way through, using
clubs to clear a path.

"What is the matter?" I asked of a businessman on the edge of
the crowd.

He shrugged. "It is only two automobile drivers in an argument.
This is Ramadan, you know. We have many arguments late in the
day during Ramadan."

It also is the time of the day when violent crimes are apt to be
the most serious, such as someone shooting his mother-in-law or
settling a family quarrel with a knife. Usually these are crimes of
passion, often unplanned, the result of frayed tempers after being
without food or water since dawn. Major crime rates vary between
countries and areas in the Middle East. But where Islamic tradition
is strong, there usually is respect for authority and crime rates are
low.

"I would rather travel alone across Saudi Arabia than go through
New York's Central Park alone at night," one Aramco oil engineer
said in Dhahran, Saudi Arabia. He had been mugged in New York
on his last home leave. When he lost a pocketbook in Dammam,
Saudi Arabia, someone turned it in to the local police station. A
telephone call from the officer on duty informed him that the pocketbook, with money intact, was available any time he wanted to call
for it. No hurry, though. It still would be waiting for him when he
showed up, he was assured.

"Islam preaches the One God, Allah, the creator of heaven and
hell, the ultimate governor of the universe," said Elie A. Salem, a
slender, intense political science professor at American University of
Beirut. I had met him and several other AUB professors for lunch
at the Alumnae Club of the University, adjacent to the school's new
twenty-million-dollar medical center. Salem, a passionately involved

man in anything he does, wanted to make sure that I understood the Arab and his drives.

I knew the background of Mohammed, how he preached his monotheism in a pagan Arabia, how he ranted against "graven images," how he was forced to flee from Mecca to Medina in A.D. 622 and how he returned in triumph.

"You know, of course, that the Moslem calendar dates from that flight, or what we call the Hegira," Salem explained. I knew this too, but he is a man whose enthusiasm flows right over conversational barriers. And I did enjoy listening to him. He illustrated how quickly an Arab can slide into a theological discussion, even in a luncheon meeting planned for another purpose.

I could not imagine a group of Christian political science professors slipping into a discussion about Jesus Christ over lunch. Here in Beirut, it seemed natural. Salem emphasized that three forces govern attitudes and actions of Arabs: their belief in Islam, the Arabic language, and Arabic culture.

"You must understand our religion," he said. "Islam provided Mohammed with the ideology and the movement to unify the tribes of Arabia, to make a nation out of them."

Few religions blossomed as fast as did the religion of Mohammed. It took centuries for Christianity to spread through the Roman world, and into those mysterious, long-forgotten kingdoms of the East where Nestor Christians spread the gospel. In a century, Islam nearly conquered the entire world.

One of the first cities conquered was Damascus, the city visited by Mohammed as a young boy with an uncle's camel caravan. The city is cradled in an oasis created by a river that sweeps down from the mountains to the west, only to disappear in the desert sands to the east. Minarets of mosques rise on wide squares where truck traffic outweighs the automobile. Dusty trees line boulevards that lead to the walled old city in the community's heart. In that old city are miles of narrow, winding streets, lined with bazaars always crowded with people.

This is the city where St. Paul was converted, where he had to escape over the wall as Roman soldiers sought his life. It was the home of the Omayyad dynasty, which represented that first flowering of Arab culture less than a century after Mohammed's passing. It was near the Omayyad Mosque, with its arched colonnades, that I felt the depth of Islam's hold over the people.

I had been wandering in the maze of narrow streets in the souk looking for Saladin's Tomb. A tunnel from one street led past a jewelry shop. I turned into it, felt my way in semi-darkness into a tiny court. Sunlight lit the small square where twenty or so boys sat on palm mats before an old Mullah. All sat cross-legged, the turbaned holy man holding a copy of the Koran. Shaven-headed youngsters listened, faces eager, eyes intent on lips of the gray-bearded patriarch as he read a verse from the Koran.

Then, in sing-song style they repeated the verse. Discipline was so strong that boys held their seats even after sighting me. The old Mullah turned, nodded with a smile. Then, he resumed the lesson.

These boys would know the Koran by heart before they reached their teens. Their education would be based upon it, and its laws would become a part of them, to leave them with a sense of guilt should they fall away from the rules of Mohammed in later life. I knew some of those verses myself, though not in their original archaic classic Arabic.

Lo! Those who devour the wealth of orphans wrongfully, they do but swallow fire into their bellies, and they will be exposed to burning flame.

And if the debtor is in straitened circumstances, then let there be postponement to his payments; and that ye remit the debt as almsgiving would be better for you if ye did but know.

Man can have nothing but what he strives for.

Such were the substances of Mohammed's sermons in market places and street corners as he converted the pagans of Arabia to Islam.

In old Jerusalem I once asked an Arab friend why the city was holy to Moslems. I knew that Islam recognizes Abraham, Moses, and other Old Testament prophets, and includes Jesus Christ as a prophet too. But why, I asked, was Jerusalem such a holy city along with Mecca and Medina for Moslems?

We were in the parking lot of the Inter-Continental Hotel on the Mount of the Ascension. Morning sunshine glinted on the golden Dome of the Rock Mosque below us, while the limestone walls which encircled the old city were just beginning to lose their early-morning pink. Behind us the Judean Hills fell away to misty distance over the Dead Sea.

"Mohammed ascended into heaven from the rock which is now covered by that dome," my friend pointed to the mosque below us.

Later that morning we drove down the steep road to the town and my friend served as a guide when we visited the spot. The story is that one night, while sleeping in Mecca, Mohammed was called to Jerusalem by the Angel Gabriel. There he found a white horse awaiting him. On it, he was carried into heaven to meet all of the preceding prophets.

Mount Moriah, site of the mosque, is also reputed to be the place where Abraham started to sacrifice his son Isaac, when God bade him desist. The bare white limestone of the original mount has been left exposed within the mosque. With shoes removed, I followed our guide into a grotto within the rock.

"When Mohammed was in heaven, he told Moses that he was establishing a rule that every Moslem should pray fifty times a day toward Jerusalem," my Arab friend whispered to me. "Moses was practical. He said that wouldn't work because nobody would pray that much. He said five times a day would be enough, and that is what Mohammed decreed."

Moslems did pray when facing Jerusalem in the early days of the religion. Then, Mohammed reported that another revelation called for them to face Mecca instead.

We departed from the Harem esh Sharif (Noble Enclosure) of the mosque via the portal overlooking the Wailing Wall. Several dozen Jews prayed facing the wall, which forms one side of the foundation of the Harem esh Sharif. My Arab friend fell silent as he walked around the space that has been cleared before the Wailing Wall. I said nothing either. I have friends among Israelis and among Arabs. I can understand the feelings of both in the Israeli-Arab confrontation.

On one of several trips to the Red Sea port, we took Middle East

Airlines from Beirut. When boarding the Boeing 707 I noticed that one of the attractive stewardesses was fingering some worry beads, the *misbaha*. I asked to see them.

"Certainly, sir. You may keep them." She handed me a string of amber beads resembling a Catholic rosary without the cross.

I remembered something I had been told. Don't ever admire an article in Arab lands or the host might give it to you. Now I felt embarrassed, for I had merely wanted to have a look at a pair of these beads that one sees everywhere in the Middle East. Businessmen, shopkeepers, taxi drivers, everyone seems to have his *misbaha*, and I never had been told about their relationship to Islam.

"I can't keep them," I protested.

"You must," she said. Then she reassured me. "We are giving a set away to everyone on this plane."

I knew Sheikh Najib Alamuddin, the American University of Beirut graduate who has built MEA into one of the best small airlines in the world. It is the sort of promotional gimmick he might have suggested.

Now that I had the string of beads, I didn't know what to do with it. My seatmate turned, smiled at me, and reached into a pocket. He took out a string of well-worn ebony beads, played with them in his fingers.

"That's all you do with them. Just that." He clicked the beads, ran them through his fingers. "See. It keeps your fingers occupied, steadies your nerves."

"Is that all?" I asked.

"That's all." He took this as a huge joke. "Many of us in this part of the world would be lost without a *misbaha*."

"Then it's sort of like reaching for a cigaret."

"Yes," he said. "Many Moslems don't smoke, and, of course, we aren't supposed to drink alcoholic beverages. The *misbaha* is a relaxer."

A pacifier, if you will. I did discover later, though, that the *misbaha* originally came from India. The Sufi sect of Moslemism adopted it to help recall the names of Allah, such as "The Merciful," "The Compassionate," "The Patient," "The Wise," "The Venerable," "The Giver of Life," "The Giver of Death," and so on.

For most Moslems, however, it has no religious significance. Catholics started using the rosary in the thirteenth century after Franks came into contact with Moslems during the Crusades. Since Moslems had been using the *misbaha* for four centuries at that time, the idea of the Catholic rosary may have come from Islam.

The Moslem calendar is one invention that was not adopted by the Western world. The year A.D. 622 was established as the beginning of the Moslem era after Mohammed's death. The Gregorian year of 1973, for example, is 1393 A.H. (After Hegira).

You cannot simply subtract 622 from the Gregorian calendar to find the Moslem year, because the Moslem year has 354 days, with eleven years of every thirty being leap years of 355 days each. The Islamic calendar is based on the lunar rather than the solar year. The 12 months of the year are each 29½ days. Thus, 100 Islamic years are equal to 97 Gregorian ones.

Since Arabic years are shorter than the Gregorian, the months are not tied to seasons. The first day of the new year comes earlier and earlier each year by reckoning on the Gregorian calendar. The Moslem year of 1387 started on April 11, 1967, of the Gregorian calendar. The year 1393 started on February 6, 1973.

A Moslem holy day such as the Feast of the Sacrifice, *Id al-Adha*, will, over a period of centuries, fall on every day of the Gregorian year. It is as if Christmas and other Gregorian holidays floated, coming in winter, spring, summer, and fall, according to the day that opened each new year.

This is much less complicated to the Moslem than it may sound to the non-Moslem.

In 1973, the holy days fell in January, a cool month in Jidda and Mecca.

"When the *Hajj* is in summer—" Abbas Sindi, the Ministry of Information executive who became our adviser, guide, confidant, and *mutawwif* (pilgrim's guide) at Jidda, did not finish his sentence. But he shook his head. We knew that temperatures rise above one hundred degrees Fahrenheit with high humidity.

"So you roast," I said.

He nodded.

We were sitting in the lounge of the Jidda Palace Hotel in the heart of the port city. The first rain in two years was underway, a torrential downpour. Since Jidda's streets have no drainage, water flooded the streets. A driveway of the hotel became a creek with the water threatening to flow into the lounge. As we sipped mint tea, porters were rolling up the carpets to save them in case the help couldn't contain the water.

Sindi relaxes easily. He thought this was as good a place as any for him to be. With a sweep of his white robes, he sat down and called for tea, pleasant and unruffled, looking a bit like Groucho Marx masquerading as an Arab. He really could not be our *mutawwif*; as non-Moslems we could not make the *Hajj*. But he saw my interest in Islam, and he needed little encouragement to talk religion as the rain poured down outside. Arabs are poor propagandists when they seek to sell their political slogans in the West. They are super-missionaries when describing their religion to any visiting foreigners.

"I come from Mecca. Islam is part of me, like my own flesh," said he. Then, he smiled wryly. "But during the *Hajj* everyone in the Information Ministry is working nearly twenty-four hours a day, and the work starts several weeks before *Hajj* and continues several weeks after."

Making the pilgrimage is one of the five major obligations of the devout Moslem. Obligations are:

> To accept belief in one God, with Mohammed as his prophet.
> To pray five times a day, facing Mecca.
> To give alms to the poor.
> To keep the fast of Ramadan.
> To make the pilgrimage to Mecca.

Obviously, every one of the 500 million Moslems cannot make the pilgrimage, perhaps because of incapacity, finances, or other personal reasons. But the *intention* must be there.

"You know also that the devout Moslem does not drink alcohol," Sindi said. He stirred the mint tea with a spoon, then added several more spoons of sugar. Highly sugared tea is the pick-me-up drink in most of the Moslem world. But most of the Moslem states do permit bars to operate for benefit of the non-Moslems, and for any fallen-away-Moslems. Saudi Arabia is tightly prohibitionist.

"How many wives do you have?" I asked Sindi. This is a question that the average Westerner seems to ask everytime he gets acquainted with a Moslem. The average Moslem regards the question with amusement. He relishes being viewed as a stud able to service several wives and concubines.

"One is enough," said Sindi. "Americans have the wrong idea about marriages in Islam. Wives are expensive. Not many men have more than one."

Polygamy, of course, has been practiced in many societies. In the hard environment of old Arabia, polygamy propagated the race and was practiced since early times. Moreover, Mohammed favored the ladies. At twenty-five he married a rich widow, Khadija, who had inherited a caravan business. Subsequently, he took other wives and fathered a huge brood. His ideas of marriage became established as law in Islam. This doesn't mean that it is compulsory to have, or that every Moslem wants, more than one wife.

Islam early split into several branches as did Christianity. Two key groups are the Sunnis and the Shiites, of which the former is much larger, some say 90 percent of all Moslems. The Shiites or Shias are found mainly in Iraq and Iran. There are also several other splinter groups, such as the Ismailis and the Druses. Within groups there are movements, usually of a reform nature, such as the Wahabis of Saudi Arabia.

Once, during the *Hajj*, we took the road to Mecca in a Chevrolet Impala with Sindi and a devout driver who was perplexed at our presence. Traffic held speed to a slow crawl. Tens of thousands of pilgrims in white robes sat in seats of taxis or limousines, or on bales of merchandise on back ends of trucks. One youth zoomed by on a motorcycle, robes flying in the wind.

A few miles from Mecca a military post bars the way. Passports are checked and no non-Moslem is allowed beyond this point. *Hajj* pilgrims will already have received their *Hajj* papers through agencies maintained in every Moslem country.

"Suppose someone pretended to be Moslem?" I asked.

Sindi shrugged. "It would not be pleasant for an imposter."

As we returned to Jidda, I noted a huge sign of painted white boulders set on a hill to form a neat Arabic script. Surely, I thought, this must be a religious inscription so located on the edge of Islam's most holy ground.

"What does that sign say?" I asked Sindi, ready to hear sacred words, perhaps from the Hadith, Mohammed's book of sayings.

Sindi squinted in the sun. "Oh, that. It says, 'Use Omo Soap.'"

The Arabic language has shaped minds ever since the time of Mohammed. It should not be surprising to find it in advertising.

When Mohammed's revelations were collected by his successors into the Koran, this became the first book in Arabic. Writers of the seventh century used the refined speech of the day, and their interpretation became accepted as authentic grammar for classic Arabic. Because the Koran was revealed in it, it became a holy language, too sacred to change, a polished, stylized idiom far removed from the colloquialisms of the street.

All languages govern cultures, shape forces within society. Some languages are more powerful than others and Arabic is a strong one. Roman Catholicism early adopted Latin and it became a binding force in the Church. English has become the lingua franca of world business, pop culture, and science. It carries Anglo-Saxon culture everywhere, much to annoyance of the French, who feel that the world would be better off with French.

Linguistic eloquence became a virtue in Islam. In the eighth and ninth centuries, when Arabs had conquered half the known world, their language was the binding force for welding empires together. It served first as the medium for absorbing cultures of Greek and Persian predecessors. Then Arabs used their classic Arabic to transform those cultures into something of their own, a truly Islamic culture.

Today, when Arabs boast of the glory of past civilizations, they usually mean that period during the rule of the Omayyad dynasty in Damascus and its successor, the Abbasid dynasty in Baghdad, when Arabs excelled in medicine, optics, astronomy, mathematics, and other sciences. Books of savants were written in Arabic and translated into Latin to become text books in Europe. Arab knowledge helped preserve culture while Europe was in its Dark Ages. Al-Baladhuri's *History of the Conquests* furthered the study of history as an intellectual pursuit. Rhazes, a great Arab physician, performed pioneer research in smallpox, detailing finds in a report written with penetrating clarity. The science of algebra was developed.

That glorious era ended with internal bickering. This was followed by invasions of Mongol hordes, Turkish conquests, intervention by Western powers, and Arab subservience.

In the nineteenth and early twentieth centuries Western powers dictated events in the Middle East. In Iraq when a guest overstays his leave, the Iraqi says, "He stays like the British." Britain occupied Egypt in 1882 for what it termed a "temporary" period, and troops were not withdrawn until 1952. When France won certain rights to represent Catholics in the Middle East from a corrupt Turkish ruler, it transformed that "right" into a political weapon for colonializing parts of the Arab world. Russia continually sought a route to a warm-water port in the south through the Middle East. Britain barred any such move by Russia.

That rivalry of Western powers existed into the oil age of the twentieth century. It provided a background for the first discovery of oil in the Middle East, a discovery that was finally to provide Middle Easterners with the strength they had lacked for centuries.

IV

Emergence of the Oil Age

Nature in its manufacture of oil pays little attention to man and his needs. Oil in the Middle East has been distributed very unequally, as were the capabilities of men. If geographic liberals existed, they would agitate for a redistribution of the oil along egalitarian lines. This is impossible without geographical alterations of the planet. Moreover there is only a slight inclination to redistribute any of the oil wealth beyond a nation's borders. Politics rather than need dictates the flow of any such largess.

Jordan and Lebanon have no oil at all. Syria and Israel have only small amounts. Egypt, with the greatest need for revenue, has some oil, but disappointing production. Kuwait and Abu Dhabi are the richest per capita nations in the world. They attained this rank because of copious oil resources coupled with small populations.

Iran, with a population of over thirty million, is the only country in the Middle East that joins a large population and much oil. It was the first nation in this part of the world to produce oil. Both Iran and Iraq had long been recognized as potential sources for it, thanks to seeps that advertised its presence even in prehistoric times. Utnapishtim, the "Noah" of the Gilgamesh legend of Sumerians, used pitch to caulk the ark that he built. The walls of Old Babylon were cemented with bitumen. Ancient physicians collected oil and attributed medicinal qualities to it.

Before the nineteenth century nobody knew how to extract oil in any quantity. Not until that century did economic necessity provide the incentive. In 1850, Scots researchers discovered how to make kerosene from coal. Overnight it became a lamp fuel, better and cheaper than whale oil or candles. As use of kerosene spread, companies investigated crude oil as another source for lamp oil. America showed how the job could be done, with Pennsylvania as its laboratory. Oil seepage in the western part of that state near Titusville had already attracted attention.

Edwin L. Drake, a drifter with a mechanical bent, suggested drilling for oil the way men drilled for water. On August 27, 1859, he proved his point with a drilling rig when he struck oil. The petroleum gushed forth like bubbling water as he said it would. A new industry was born, which burgeoned around the world so fast that all lands seemed to lie downhill for oil.

By 1900 America had a booming oil industry. Russia dominated European output. In 1902, when Spindletop, the first Texas oil well erupted near Beaumont, Texas, America solidified its position as a boisterous, young industrial power.

About that time, Persia, now called Iran, was attracting attention as an oil source. Today, flying over the sprawling petroleum refining and shipping complex of Abadan at the head of the Persian Gulf, one finds it difficult to imagine that the land below was a barren, uninhabited island at the turn of the century. Acres of silvered oil tanks gleam in the sun. Towering smoke stacks belch smoke above a maze of pipes, retorts, giant cylinders, and other paraphernalia of the largest oil refinery in the Middle East. Beyond the tanks, row on row of neat oil workers' homes stretch in military-like formation into hazy distance.

In Abadan, a young Englishman fresh from university and still wrapped in an unearned layer of superiority, crisply told us, "Iran's oil history is tied to that of British Petroleum Company." He braked the English-made Ford Cortina on the Shatt al Arab waterfront. Pickup trucks rolled along the wide boulevard, driven by Bakhtiaris, Kurds, Lurs, or other Iranians who have been trained in the apprentice schools first established by the company. The pungent smell of hydrocarbons hung over the whole area, a constant reminder of the reason for Abadan's existence. A tanker, BP on its stack, edged slowly through mud-brown waters, seeking a berth for a cargo of aviation fuel for some distant point.

"Yes. Our history has been tied to British Petroleum," a twenty-four-year-old, moustached public relations officer of National Iranian Oil Company admitted shortly after. As an afterthought, he added, "Unfortunately." He had recently returned to Iran after graduating from the University of California. He liked everything American, including its coeds, and had little good to say for the British, saying, "They exploited my country from the very first."

Prickly nationalism is common in Iran as well as in other places in the Middle East. It galls Iranians to remember how helpless Iran was in 1900 when BP existed only as a dream in the head of William Knox D'Arcy. He had made a fortune in the Australian goldfields, then returned to his native England. There he heard that Shah Muzaffar ed-din of Persia was offering oil concessions in his country. D'Arcy grabbed at the chance to put his money to work.

Russian influence was strong in northern Persia. Britain was a power in the Persian Gulf, thanks to its navy. In those days, power was measured in battleships, not in balance-of-payments terms. Britain had all the floating guns needed to garner influence in southern Persia. The sun never set on the British flag mainly because the British were busily hoisting their colors around the globe.

The Shah of Persia played one colonial power against the other, taking money from each. He was too weak to do much, other than maintain a low profile. The oil search to him meant badly needed cash for a treasury emptied by frequent wars against rebellious tribesmen. If oil were found there would be a 16 percent royalty.

Shushtar, on the Khuzestan Plain in the south, became one of the bases for that petroleum search. Hostile Bakhtiaris amid the towering ranges of the Zagros Mountains threatened operations.

"Until only a few years ago every Bakhtiari carried a rifle and it was unsafe to cross the mountains without an escort," Chahar Mohammed, a sheep raiser on the Khuzestan Plain, said. We encountered him in a mudbrick cafe on the main street of Shushtar in the heart of this region, where oil men first entered the country. Shushtar is a sandy village of square houses, dilapidated mosques falling into ruin and of numerous goats. Children played along the bank of the Karun River below the plateau on which the village sits.

"Now only half of the men seem to carry rifles," I said. We had finished our breakfast of flat bread, tea, boiled eggs, and yogurt, without asking any questions about the cleanliness of the kitchen. Chahar, who had learned a little English when working at the Abadan refinery, introduced himself, then joined our table at my invitation. We found him to be much more voluble when speaking through our interpreter than when testing his seldom-used English.

Chahar laughed. "They like to hunt," said he. He was medium-sized, of wiry build, a felt skullcap on head, the lower half of his body swathed in baggy pantaloons. The sun had darkened his craggy features and had drawn prune wrinkles about his brown eyes. Obviously, he was a man of the outdoors, one who perhaps knew how to use a rifle himself.

"Do they hunt animals or men?" I asked in jest. Under Shah Pahlevi's firm rule, Iran now is a stable country, with tribes under firm control. This wasn't so, though, when oil men first entered the country. And Shushtar claims to be the hottest inhabited village in the world in summer, when temperatures of 120 degrees Fahrenheit are common.

Bad luck plagued the oil search from the first. The operation hit dry hole after dry hole. Oil geology then was in its infancy. Nobody knew enough about the basic geological structures to do much more than depend upon signs of oil seepage for locating drilling rigs.

When oil was found, it amounted only to a trickle. Company funds ran low. In 1905, D'Arcy sought financing from Burmah Oil Company, a Scottish-based company with operations in Burma. Burmah and D'Arcy joined forces for what was to become the Anglo-Persian Oil Company.

At first, the new joint venture did not fare any better than its predecessor. There were more dry holes. Tribes in the area harassed drillers. The undeveloped country added to transportation costs, and funds again approached exhaustion.

Company officials in London resigned themselves to collapse of the operation. Then, on May 26, 1908, drillers struck oil at a place called Masjid-i-Salaman (Mosque of Solomon), the scene of an ancient temple. Oil gushed forth in a torrent.

On that day, the Middle East's oil industry was born. The birth came at a time when navies of the world were realizing that oil might be the fuel of the future for vessels of war. Oil was developing a political significance, which it has retained to this day.

England's Winston Churchill pressed for British government participation in D'Arcy's company to protect the Royal Navy's fuel supplies. In 1914, Britain purchased just over 50 percent of the company's equity, a percentage that currently is just under 50 percent. In 1954, the company changed its name to British Petroleum Company, Ltd.

In 1973, BP still had an enormous interest in Iranian oil. It held 40 percent of Iranian Oil Participants, Ltd., the consortium of fourteen oil companies which was the largest producer in the country. When Iran took over its oil operations on August 3, 1973, members of the consortium drew up supply contracts with Iran according to the percentage of interests held in that consortium.

Mesopotamia, or Iraq as it now is known, was recognized as another potential source for oil. Before World War I it belonged to a Turkish empire that had grown weak and impotent over the centuries. Prior to the war, British and German interests established the Turkish Petroleum Company to seek oil in Mesopotamia. Because of his efforts in forming this company, Calouste Gulbenkian, an Armenian from Constantinople, was given a 5 percent interest. Henceforth he was to be known as "Mr. Five Percent."

The world war shelved that company. When it ended, France demanded and received the 25 percent interest of defeated Germany. Subsequently, Compagnie Française des Pétroles was created to acquire these shares. Eventually, the French government took 35 percent of the equity and 40 percent of voting shares in CFP, another indication of the interest governments were displaying in petroleum even then.

Hope for oil in Mesopotamia played a part in British repudiation of an independence promise to Arabs for helping the Allies in World War I. Immediately after that war, France and Britain split control of the Mediterranean-to-the-Persian Gulf area. Britain assumed a mandate of Palestine, established the Hashemite Abdullah as king of newly created Transjordan, and placed his brother Faisal in Baghdad as king of the new kingdom of Iraq. France took control of what is now Syria and Lebanon.

That repudiation of Arab hopes planted seeds of bitterness that have germinated. This is reflected in obsessive Arab fears about colonialism and neocolonialism and it conditions some of their bargaining with Western oil companies.

"Every true Arab nationalist regarded the Hashemites as turncoat intruders, the stooges of the British," one aide of General Abd al-Karim Kassem said shortly after the 1958 revolution had ousted the dynasty in Iraq. King Faisal II was murdered in that coup d'état. Kassem became Iraq's dictator.

At that particular time I had requested authority to visit the Kirkuk oil fields to the north, where Iraq's first oil was discovered. Permission was rejected on the grounds that Kurdish unrest made it too dangerous to travel.

"What they really mean is that Americans aren't popular here," one Iraq Petroleum Company official said. President Dwight D. Eisenhower had ordered the landing of American troops in Lebanon, and America was catching some of the backlash of the anti-British sentiment, which has always been strong among nationalists of Iraq. American support for Israel, of course, has added to that residue of ill will for Americans in the Arab world.

In the period immediately after World War I there was commercial bitterness between the United States and Britain, too, in the Middle East. America viewed British intrigues in the area as a geopolitical maneuver to gain oil for itself, perhaps at America's expense.

The United States emerged from World War I not only firmly entrenched as the world's largest oil producer, but as the world's largest consumer, too. The term "energy gap" had not been coined yet. Nevertheless, in 1920, some economists, congressmen, and oil men feared that America soon might deplete its oil at the pace of consumption then maintained (about 3 percent of the early 1973 rate of consumption). Now the British seemed intent upon capturing control of an area that oil men said contained much petroleum. Nobody had any idea of the amount involved, or British-American-French intrigues might have become more querulous.

In 1921, Herbert Hoover, then Secretary of Commerce in the Harding Administration, persuaded seven large American companies to form a consortium to work together in the Middle East— Standard Oil Company of New Jersey (now Exxon Corporation),

Standard Oil Company of New York (which evolved into Mobiloil Corporation), Gulf Oil Corporation, Texas Oil Company (now Texaco, Inc.), Sinclair Oil Company, Atlantic Oil Company, and Mexican Oil Company. That consortium, formed mainly to hoist the American flag in the Middle East, became the Near East Development Corporation.

American diplomatic pressure persuaded the British to open the door to this consortium. It was allowed to take 23.75% of the equity in Turkish Petroleum Company.

Britain and France smugly planned a gentlemanly cartel to control Middle East oil. Their companies, Holland's Shell, and any American participants would work together. Outsiders would be kept out. Those on the inside could watch each other closely so that no one country or company obtained an advantage over others.

At this, American companies rebelled. They wanted an open-door policy in the Middle East even if this meant fierce competition. They felt they could manage as well and perhaps better than anybody else in a competitive climate. They had technology, the trained manpower, the necessary financing, and the huge markets.

But France insisted that a cartel must be arranged. It backed a proposal of "Five Percent" Gulbenkian, the wily oil wheeler-dealer who often was in the middle of oil negotiations though he never produced a barrel of oil in his life. A red line was drawn on a map, encircling the old Turkish empire. Within this area companies in Turkish Petroleum would not act individually. They would cling together as a consortium, each company retaining the percentage originally allocated.

France promoted the cartel arrangement harder than did anybody else. For nearly a half century since that time it has cleverly used diplomacy to protect its interests and to promote its philosophies, often at the expense of America, whether the topic involved the United States dollar, Vietnam, a separate nuclear force, the North Atlantic Treaty Organization, or Quebec autonomy.

France wanted an equal share of Middle East oil even though it was far behind America and Britain in oil technology, financing, and development. It realized that in any competitive free-for-all it would be clobbered. With the cartel arrangement, it would be guaranteed 23.75 percent of the area's oil.

France seemingly won its point. The Red Line Agreement was accepted by all parties in October 1927. Still, the agreement had little meaning to American companies, except for Gulf Oil Corporation, as subsequent developments showed. Even as the new consortium became Iraq Petroleum Company, Ltd. (IPC), several American members dropped out.

"It would have been suicidal for America to have accepted the Red Line Agreement," one American diplomat in an embassy in the Middle East said during a discussion of this topic. He stood up, strode to a map of the area hanging on a wall. "Look." He encircled the Arabian Peninsula and the Euphrates Valley with a sweep of his hand. "That includes all of the oil production of the Middle East except for that of Iran and Egypt. If we had let them put us in the Red Line straitjacket, America would have been held to twenty-three and three quarters percent of the oil in that area. It was a cartel arrangement by Europeans to bottle up American companies."

IPC proved profitable for the companies that did remain in the group. On October 14, 1927, oil was found near Kirkuk in the northern part of Iraq.

"It proved to be one of the most remarkable strikes ever made," one veteran IPC oil man said in his Baghdad office. Permission to visit the area was easier to obtain on this particular trip to Iraq, just before Iraq nationalized those fields in June 1972.

A photograph of that first well, Baba Gurgur No. 1, shows it discharging two plumes of oil 140 feet into the air at an 80,000 barrels-a-day rate. This oil flooded the countryside.

An American, H. A. Winger, was the tool pusher on the drilling rig. Another photograph shows him standing by an uncapped well, a pith helmet on head, husky and broad-shouldered, looking as if he could handle himself in any kind of company.

His report details how the drill bit was being pulled to clean it when the well exploded its millions of pounds of pressure. With a rumble and a hiss, oil shot upward, blowing tools out of the way. Fortunately, working men scattered at the noise beneath their feet. Winger hurriedly doused every flame in the area. This was fortunate too, for natural gas poured out with the oil.

It took ten days to bring this gusher under control. Meanwhile, a lake of oil accumulated in the countryside. Men worked feverishly to cap the well, all the while fearing that a spark might ignite the gas and create a raging inferno.

Capping operations were complicated because the Eternal Fire burns only a mile and a half away. This is a seepage of natural gas that ignited in prehistory and has been burning ever since. Baba Gurgur means "eternal fire," a salute to those hissing flames that may have been the "fiery furnace" described in the Bible. A work force of several hundred Iraqis kept oil from the Eternal Fire. Everything was brought under control and Iraq became a major petroleum producer.

There is little to see at Baba Gurgur No. 1 today. But the Eternal Fire still burns on a rocky plain. Flames cover an area of about ten yards across. It is like a grass fire that always remains in one place, but much hotter.

IPC's holdings developed along two lines. It and an affiliate, Mosul Petroleum Company, operated fields near Kirkuk and Mosul. Oil was piped across Syria to outlets on the Mediterranean at Haifa, Palestine, and at Tripoli, Lebanon. After Israeli independence in 1948, the Haifa pipeline was abandoned. An outlet in Banias, Syria, bolstered that at Tripoli.

In June 1972, Iraq nationalized the northern fields of Iraq Petroleum Company. Subsequently, Syria nationalized that part of the pipeline on its territory.

Another IPC affiliate, Basrah Petroleum Company, developed fields in the delta region of the Euphrates and Tigris rivers. Its oil is shipped to market through the port of Khor al Amaya, on the Persian Gulf.

The Near East Development Corporation, America's foothold in the Mideast, evolved into an equal Exxon and Mobiloil partnership. In October, 1973, Iraq nationalized those holdings as an anti-American gesture because of U.S. aid to Israel. Gulf Oil, one of the original participants, had sold its holdings early to the partners. Gulf, a company put together by the Mellons of Pittsburgh, however, was to be both the ironic victim and beneficiary of a set of circumstances stemming from the infamous Red Line Agreement.

Now the stage was set for the launching of an Arab oil industry on the Persian Gulf.

V

The Oil in Aladdin's Lamp

The Standard Oil Company of California (Socal) has been so successful over the past three decades that it is difficult to imagine that it once was a jinx company internationally. While everybody else seemed to discover oil abroad, Socal found nothing but dry holes. Oil can be like that sometimes, elusive, hard to find, often located in godforsaken places when it is found.

This is why enormous capital expenditures are necessary for discovering and developing new petroleum fields. This also is why the poverty-stricken countries of the Middle East had to accept help from outside to develop their oil through the first half of the twentieth century and well into the second half. The early history of Bahraini, Saudi, and Kuwaiti oil is interwoven with Gulf Oil Corporation's ironic story and with the miracle tale of jinxed Standard Oil Company of California, which overcame its hoodoo and seemed to find Aladdin's Lamp in the Middle East.

Long before that, however, sheikhdoms of the Persian Gulf were accepting the protection of Britain, anxious for the welcome shade of the billowing sails on British warships.

Modern Kuwait gives no clues at all to its early history. In area this little sheikhdom is a trifle larger than Connecticut, a table-flat desert land where people drink seawater (distilled) and collect crude oil from the wells that dot tawny sands. Most of the 740,000 population resides in the city of Kuwait or in its sprawling suburbs.

Kuwait is a city of wide four- or six-lane highways laid among a myriad of apartment and business buildings strewn haphazardly beside a listless Persian Gulf. There are a few mudbrick houses left from pre-oil days. Most of the city is composed of buildings with daring designs, as if a group of futuristic architects were given blank checks to turn their dreams into reality. Cubes and blocks of concrete and steel are interlaced with balconies and terraces. The Kuwait-Sheraton Hotel is a high-rise structure of glass and steel with an ornate lobby that could serve as a reception hall for an Oriental prince. The competing Kuwait-Hilton is by the sea, across the street from the American Embassy. It is a towering and luxurious establishment with smart shops in the foyer.

On Fahad al-Salim Street, shops, banks and multi-storied offices hum with activity. Modern supermarkets offer everything from cans of Heinz baked beans to Hershey chocolate bars. A forest of television antennas sprout above homes and apartment buildings, advertising the presence of some of the seventy-thousand sets in use.

"Annual attendance at our movie cinemas totals about two million people," an executive of Kuwait Cinema Company explained. I had been curious about Kuwaiti attitudes toward movie houses after noting that they were unlawful in Saudi Arabia. Kuwait Cinema operates all eight theaters in the country, six of them indoor and two outdoor.

"And you have no religious problems?" I asked.

"None at all. We Kuwaiti are a sophisticated people," said he. "We have the highest literacy rate in the Arab world."

Kuwait is tucked into the upper end of the gulf between Saudi Arabia and Iraq. It developed into a distinctive entity after 1710 when the al Sabah clan of the Bedouin tribe of Atib moved there from the Arabian interior. By 1756 the ruling Sabahs were recognized by Turkey as rulers of an independent sheikhdom.

It was a perilous independence in a pirate-infested Persian Gulf, with marauding tribes on its shores. Kuwait looked to the sea. It became a pearling center, and the home of intrepid mariners who sailed their dhows into far ports of the Moslem world. Few people are better than the Middle Easterner at adapting to his environment whether it be a barren desert wadi, a peninsula thrusting into the sea, or a rich city like Beirut.

"And man's ability to adapt to the circumstances and conditions around him is perhaps the best measure of his intelligence," one anthropologist at the American University of Beirut said, when discussing the Arab.

In 1899, Kuwait eagerly accepted British protection. As part of that deal, Sheikh Mubarak of Kuwait agreed on January 23, 1899 that he would never give or lease any territory or concession to any other foreign government without the consent of the British government.

In those days of empire, Britain used to think that its actions were for the good of native regimes. If its oil industry or its traders obtained benefits from these maneuvers, this was a bonus for colonial goodness. Such bonuses abounded before World War II. The Kuwait agreement was to be interpreted later by Britain to favor its own oil companies first for any oil concession Kuwait might grant.

Saudi Arabia, which was officially given this name only in 1932, was a tougher nut for Britain's diplomatic nutcrackers. King Abd al-Aziz ibn Abd ar-Rahman al-Faisal al-Saud was a hard-bitten desert warrior who united his country through battle after battle with untamed desert tribes. He did not need or want British protection, and made that plain in 1923 when he denied an oil concession to Anglo-Persian Oil Company, the 50 percent British government-owned company. Later the king did grant a concession to Eastern and General Syndicate, another British outfit. But Eastern and General had no connection with the British government. In fact, its founder, New Zealander Frank Holmes, frequently was on that discreet blacklist that Britain's Foreign Service maintained for those who refused to play the game by British rules.

Holmes, a short, heavy-set man with a gift of blarney, was one of those promotors who seem to abound in international oil, the go-between who brings companies, money, and concessions together. Such men may be catalysts for gigantic undertakings. They wheel and deal, often with empty pockets, always with glowing tales about the concessions under their control.

The New Zealander believed that oil existed in Bahrain and in the Arabian Peninsula. Though he faced competition from the large Anglo-Persian Oil Company and a British government that did not think too much of him, Holmes obtained the oil concessions for both Bahrain and for what was to become Saudi Arabia.

His own company lacked money to develop those concessions. He turned to the majors, using his charm as a salesman to sell the concessions. But Iraq was attracting most companies. Holmes got nowhere until he approached Gulf Oil Corporation. Gulf took an option on the Bahrain concession, dispatched a geologist there in 1927, and then tried to persuade its Red Line partners to join in a Bahrain venture. The partners not only turned down the proposal; they warned Gulf that it must adhere strictly to the Red Line prohibition against unilateral operations within the prescribed area.

Rather than antagonize Red Line partners, Gulf withdrew from Bahrain. It persuaded Standard Oil Company of California, an outsider to the agreement, to take its option. Thus, Gulf lost a chance at Bahrain's oil and perhaps also at the far richer fields of Saudi Arabia. It later proved only a small step for Socal to go from Bahrain to nearby Arabia.

Britain regarded the Middle East as its own preserve and tried unsuccessfully to block entrance of this maverick Standard Oil of California into the area. After long haggling, Socal established its subsidiary, Bahrain Petroleum Co., Ltd., on Bahrain. A British miscalculation helped. When British oil geologists decided that Bahrain had little, if any, oil, opposition of the British government to an American intruder faded.

Socal had established a reputation as a jinx company in international oil. Though a major American mainland producer, it had drilled thirty-seven dry holes in a row in six different countries. Moreover, it had conducted explorations in a dozen additional countries without doing anything but spend literally millions of dollars of stockholders' money.

Now it was ready to gamble stockholders' money again in Bahrain. Oil is a business for the informed gambler, the man who has some information but not enough to guarantee a sure thing on his investment. Profits may be enormous with a strike, losses may be enormous with dry holes. Often critics of the oil industry look only at those profits, overlooking the fact that literally thousands of petroleum companies have gone broke over the years when gambles failed to pay off.

Bahrain consists of one main island thirty by ten miles, and thirty-two smaller ones. It has a history as ancient as any in this part of the world. Its port of Dilmun dominated the gulf trade in the time of the Sumerians, three thousand years before Christ. The curious grave mounds of the long-forgotten Dilmun civilization, the largest-known prehistoric cemetery in the world, are an impressive sight. For mile after mile these tombs, over 100,000 of them, form blisters on the arid landscape, like some giant gamesboard waiting for an unknown game.

"In the seventh century, Bahrain was one of the first areas outside the Arabian Peninsula to accept Islam," said Ibraham Arayad, a lean and bony teacher in an Isa Town school. He said this proudly, the way many Moslems do when reporting something about their religion. His father had been a roustabout with Bahrain Petroleum Company (Bapco) in its early days, working with a crew of Texans.

Socal buried its jinx on Bahrain. With its very first well, it struck oil, in early June 1932. Within a few years, production reached twenty thousand barrels a day, a respectable total in those days. In 1936, Socal took the Texas Company (Texaco) as a fifty-fifty partner in a deal negotiated because of Texas Company's marketing facilities in the Far East.

"Then we had two American oil companies instead of one," said Arayad. "Our people quit pearl fishing and became oil workers."

Perhaps it was at the right time. The Japanese were starting to grow cultured pearls and Persian Gulf pearl divers were facing hard times. I said as much.

Arayad nodded. "But if it hadn't been oil, we Bahraini would have found something else. All through history we have adapted to whatever circumstance lay ahead of us."

"We Bahraini!" I repeated the term to myself. He was an Arab. But he was a Bahraini first, proud of his 30-mile-long homeland which has only 200,000 people.

He displayed more pride when serving as an impromptu guide driving his Vauxhall sedan about the streets of Isa Town, the government housing project that is a model city for the whole Gulf. Teacher Arayad wore his London tailored suit as if he had never worn anything else. Casual Western dress was evident on the traffic-free shopping mall of the town. The *keffiyeh* or *gutra* was in evidence, too, as was the *dishdashah*, that long, flowing white robe that permits free circulation of air about the body.

The Arab clings to his traditional style of dress because it is the most practical for the hot, dry climate of this area. That head scarf effectively shields his head from the sun. It is held in place by a black or white wool ring called an *agal*. Originally, the *agal* served as a camel hobble. Today it is simply a part of the Arab's wardrobe.

The Arab wears a small skullcap under the head scarf. When sand blows across the desert, he wraps that scarf around his face and breathes through its folds. When the air turns cold—and it can be close to freezing some nights in the desert—robes and head scarf help keep him warm. In chill periods, a heavy wool cloak, or *aba*, shrouds his body.

"I dress Arab style at home, and Western style at school," said Arayad.

"Why not Arab style all the time?" I asked. He was so pleasant, so eager to please, so proud of his tiny country that I hated to ask the question. It could have been embarrassing, for in some developing nations a pronounced inferiority complex drives natives to adopt Western-style dress as a uniform to show superiority over brother natives. Yet they may feel inferior to the bossy Europeans, something they may deny angrily if asked about it.

Still, anyone seeking information sometimes must ask penetrating questions. Arayad fielded the question easily, lips creasing in a smile beneath his bony nose. He was too extrovertic to be embarrassed.

"My students pay more attention to me when I am dressed in Western style," he said.

"Why is that?" I inquired.

We rounded a corner of a wide boulevard lined with neat, square houses set behind compound walls that enclosed green gardens. Boys played British soccer on a sandy field. Autos were parked near several of the houses. People on streets had that prosperous air of folk who know from whence their next pay packet will arrive.

"I suppose," he said, "it is because our schools have a British curriculum. Look." He pointed to the cluster of buildings belonging to the Gulf Technical College on the edge of town. "That's the best technical school in the Gulf. All classes are taught in English and students take British level exams. I think our students have come to feel that if you want the best education it must be Western style,

British, or perhaps American. We never have been orientated toward France as they have been in Lebanon and Syria."

"So when you wear a Western suit you seem to be a better teacher?"

"I am," he said with a nervous laugh. "Funny thing. When I am in western clothes, I find it easier to think like an Englishman. When I am in Arab clothes I am much more of an Arab. My Koranic education seems to become much stronger."

"That may be why Nehru of India and Mao of China led a return to native styles of dress in their countries," I said. "They wanted to emphasize their own cultures rather than that of the West."

For the first time, Arayad looked troubled. He swung the Vauxhall onto the main road leading back to Manama, the capital, and the plush Gulf Hotel. "You may be right," he said, "and that bothers me. We can't return to the culture of the desert. We must modernize, and please note that I didn't say 'Westernize.' Yet our Koranic culture is a stabilizing factor that should not be left behind."

He fell silent, his brow furrowed in thought. When we resumed our conversation, it concerned the history of the island, and its checkered career following introduction of Islam. It had thrived as an independent sheikhdom, became a Portuguese colony from 1522 to 1602, a Persian colony for a while after that, then became independent again. In 1782, the Al-Khalifa dynasty, which still rules, was established.

"Our dynasty goes back to the time of Britain's George the Third," said Arayad.

"And to the American Revolution," I said.

"We know it as the American Rebellion," he said. "We still get many of the books for our schools from Britain, you know."

Later, at a party hosted by Arayad, I had an opportunity to delve into the love-hate relationship that often seems to exist between a Middle Easterner and the foreigners who helped develop the oil of the area. Arayad owns a roomy, two-story house that looks like a box on the outside, but which is surprisingly spacious within. Persian carpets lay on the floor in the large living room, breaking the monotony of bare plaster walls. Arabs generally do not feel that a bare space on a wall must be filled with a lithograph or painting. The only wall decoration was a Koranic inscription on a green background in Arabic. Several red plush divans and chairs provided seating. In a village in the Middle East you might find yourself sitting cross-legged on the floor. Not so in the house of any town-bred Middle Easterner of today.

The guests included several of Arayad's students, a physics instructor from the Gulf Technical College, and a couple of administrative employees of Bahrain Petroleum Co. (Bapco) and wives. The modern Arab woman is beginning to break out of the shell that Islam has kept around her for many centuries. In villages, and among less-educated, low-income groups, women still are very isolated from society, masculine or feminine. The married woman of this society is likely to see only relatives, living a life segregated from all males except for her husband, his father, and his brothers.

Women may have much more freedom, at least among the educated, than they had in the old days, but they still are second-class citizens in many ways. In Arab lands, a husband's brother, for instance, may dictate how the wife shall behave if the husband is not around. Her father-in-law may lay down strictures that she will be expected to obey. She may be expected to be on the streets only when accompanied by a male of the family. In effect, she finds herself married to her husband's entire family, not just to her husband insofar as social relationships are concerned.

One pretty, dark-eyed girl in the party shook her head when I asked her if she ever wore a veil. "I never did," she said. "Even my mother stopped wearing one before I was born." She had spent three years in England at a select girls' school, and she admitted: "It would be impossible for me to return to the old ways, *purdah* and all that."

Refreshments consisted of orange juice and various soft drinks for those who lived by strict Koranic rules, but the host served Scotch and gin to those who favored such drinks, and I noted that he took a gin and tonic himself. When I asked Arayad about it, he said, "When Mohammed talked against alcohol I think he was condemning any drinking to excess. Basically, our religion preaches moderation in all things. So there's nothing wrong about a moderate amount of alcohol."

He rolled his drink around in his glass and stared wryly at it as he added, "Of course, I wouldn't want to see liquor advertised on television or radio, or anywhere else for that matter."

I have noticed this ambivalence in the Middle Easterner, in many ways, over the years of traveling through this part of the world. Liquor may be taboo to the orthodox Moslem, yet he may rationalize as he takes an occasional drink.

Ambivalence is evident in other ways. Arabs hospitably welcome strangers into households, yet the European or American may sense a resentment when visiting crowded souks. Modern Arab women claim that they now are free; yet at parties they migrate like chickens to their own corner as men discuss topics considered to be far above mere women. Foreign oil companies will be damned as exploiters of national wealth, yet the Middle Easterner voicing this complaint may be a graduate of an oil company–financed school, drawing a comfortable wage or salary from that same company.

Many Arabs are thin-skinned, ready to take offense at any slight, real or imaginary. They admire Western culture and technology, especially American. In their admiration they sometimes overlook the strengths of their own culture, seeing only its weaknesses in comparison with that of the West. They fear being regarded as inferior even as they suffer from an inferiority complex. So they criticize the West even as they adopt Western customs. Western institutions such as the YMCA becomes the YMMA (Young Muslim Men's Association) while the Red Cross becomes the Red Crescent. They like Western movies, phonograph records, tapes, and books. Yet they remain Arab.

"Our culture is a mixture of the old and the new," one of the students said. He was in his twenties, older than the students you might encounter in the United States. But this is common in most Middle Eastern nations, for the student might spend his first ten years in a Bedouin tent receiving no schooling at all. "Nihad Ghadri, one of our writers, says of Saudi Arabia that it is the only country which has its past reaching right up to its present. I think there are other countries in this area like that too."

"Including Iran?" I asked.

"Even more so," somebody interjected quickly, and everybody laughed. Feelings between Bahrainis and Iranians are edgy, just as they are between Iraqis and Iranians. Until recently, Iran claimed Bahrain and some Bahrainis uneasily wondered if the decline of British power in the Persian Gulf meant that this island might be finding itself again under Iranian overlordship.

"You see"—Arayad was passing a tray of kebabs and a platter of humus, that tasty mash made from chick-peas—"there is a duality about the Arab character because in many ways we have one foot in the fifteenth century and the other in the twentieth."

"Not here in Bahrain, though," the student said, quickly. Everyone grinned and nodded, except for Arayad.

"Even here," he said. "You will find this to be true where countries are rapidly advancing. There is a struggle between the old and the new, and some of the old and the new is in all of us."

"That's right," one of the oil-company workers, an accountant with an American University of Beirut degree, said. "Two years ago I made the *Hajj*. I went more out of curiosity than anything else. Then in Mecca I felt something, as if I were returning to the faith and innocence of my childhood, that faith that somehow everything is going to be all right in the end. I realized then how strong Islam's hold is on me." He smiled ironically as he glanced at the nearly finished glass of Scotch on the rocks in his hand, adding, "Even if I may not be a very strict Sunni."

I nodded, understanding what he and others were trying to say. The Middle East is in a period of transition. Old ways are intermingled with the new. Nationalism is rising, clashing with traditional chains of command, which often were established or supported by outside influences. A slowly emerging middle class tastes fruits of progress and wants more.

"Sometimes when we seem to be ambivalent we are merely reflecting two different sides of our backgrounds," explained Arayad.

Or three sides, or four, or five. Nowhere is this more true than in Middle Easterner attitudes toward oil and the companies that lift it from Arab and Iranian soil. Certainly there may be appreciation for the paternal social-welfare programs of oil companies for their employees. Only a fool refuses largess earned with his own sweat. Moreover, the Koran stresses the necessity for giving alms. Therefore the receiver is really making it possible for the giver to live a good Koranic life.

There may be respect for the production and merchandising expertise of those companies. The Arab is not a good organizer. This is especially true when commands filter down through the middle and lower levels of a bureaucracy, a gallon of energy and finance going with them at the top, a drop-by-drop trickle emerging at the bottom. Thus the Middle Easterner admires efficiency in others.

Middle Easterners are superb bargainers, so their respect for merchandising know-how represents the approval of one sharp camel-trader for the bargaining instincts of another. But there is deep resentment about any foreign control of the oil flowing from Middle Eastern lands. This grates on the chauvinistic pride of the Middle Easterner, whether he be Arab or Iranian.

"Never again will a huge foreign combine be given untrammeled managerial authority over a vital sector of our economy," the Shah of Iran stated when defining his program for national control of Iran's oil resources.

In Beirut, Lebanon, Abdullah Tariki, the University of Texas–educated Saudi who has been a persistent gadfly to international oil companies, bluntly said, "The international oil companies have been exploiting our oil resources for their benefit, not ours." Tariki served as Saudi Arabia's oil minister from 1960 to 1962. Since that time he has been an independent oil consultant, advising the Egyptian, Algerian, and Kuwaiti governments, and others on oil matters. He is of slender build with dark, wavy hair, handsome as a matinee idol, usually nattily clad in a well-tailored suit. Cockily, he can rant against major oil companies an hour at a stretch without provocation.

There are intelligent ministers and administrators in Middle East governments, however, who realize that it takes trained manpower and capable managers to operate any oil industry. And the oil companies do provide that talent. The Middle East has not yet reached the point where it can dispense entirely with the expatriate, and the companies that employ them. Slowly and steadily, domestic labor is being advanced to higher and higher jobs, and slowly and steadily, Arab countries are gaining more control over their oil industries.

Bahrain is in a class by itself when it comes to oil. A refinery that subsequently developed into the second largest in the Middle East was established in the 1930s. Unlike other countries that export crude oil for processing elsewhere, Bahrain exports only finished products. It produces crude at about 75,000 barrels a day whereas the refinery has a capacity of about 250,000 b/d, using Saudi Arab crude for most of its stocks.

"The Bahraini are easy people to deal with," says David Maloney, Bapco's public-relations manager. "Ministers know oil. They are sophisticated. They understand our problems as we understand theirs."

Bahrain cannot afford to make radical changes in oil matters. Its petroleum production already has peaked, and it is looking beyond oil to that day when it will have to depend upon other industries for supporting its people. Such industries may require huge dollops of foreign capital, and that will not be attracted by radicalism. Bahrain obtains the going price for oil, anyway, without being the pacesetter.

One pacesetter is Kuwait, which produces forty times more oil than does Bahrain, and which could produce sixty times as much. Instead it has frozen production at under three million b/d, and linked output to settlement of Arab-Israeli differences. This is bad news for the Gulf Oil Corp. and British Petroleum Co., Ltd., equal partners in Kuwait Oil Company. The companies had been adversaries, only to become friends in one of the richest oil fields in existence.

Gulf Oil, after dispensing with its Iraq interests, felt no longer bound by the Red Line Agreement. Its officials in Pittsburgh scanned maps of the Persian Gulf, studied geological reports, read the memoranda of exploration engineers, and in the 1930's decided that Kuwait merited exploration for oil.

Immediately, Gulf Oil encountered British government opposition. Britain resented the encroachment of American companies into the Middle East. It wanted Kuwait and all the other sheikhdoms on the Persian Gulf to remain as British preserves. What is the point of gunboat diplomacy if you cannot reap its fruits for domestic companies?

In London, American Ambassador Andrew Mellon, one-time president of Gulf Oil Corporation, eloquently argued the case for Gulf and the American presence in the Middle East. The squabble ended in compromise with the establishment of jointly owned Kuwait Oil Company. Gulf took 50 per cent of the equity, Anglo-Persian, the British Petroleum Company of today, took the other 50 percent. And it proved to be a profitable partnership for both companies.

In February 1938, the new company struck oil at Burgan on desert sands, but World War II postponed development. It was not until June 30, 1946, that the first shipment of oil left Kuwait aboard the tanker *British Fusilier*.

By that time, Saudi Arabia, too, had appeared on the oil scene. After discovery and development of its oil on Bahrain, Standard Oil Company of California focused attention on the Arabian mainland. Holmes, the promoter who had helped Socal in Bahrain, talked volubly about the potential of the al-Hasa area in Arabia's Eastern Province. In 1933, Socal opened negotiations in Jidda with the government of King Abd al-Aziz. Immediately it found itself in competition with a British delegation from IPC. Consortium members were thoroughly alarmed at the possibility of Socal gaining a foothold in Arabia. Britain believed that Americans should stay out, leaving the peninsula to the British.

To the king, IPC seemed to be a rich and powerful consortium more prone to dictate than to ask. Socal was a newcomer that had proved itself on Bahrain as a professional outfit with no obnoxious political ties. Thus Socal won the concession in May 1933. In 1936, a half-interest was sold to the Texas Co. (Texaco) to bring the latter's formidable marketing facilities into play should any oil be discovered. The place to learn the early history of Saudi oil is in Dhahran, the Arabian American Oil Company's (Aramco) paternal township on desert sands in the Eastern Province. Aramco became the production company for Socal and its later partners.

Today, Dhahran is a city of wide, curving streets with palm boulevards. Ranch-type houses sit behind low walls that form open-air patios around residences. Shade trees tower over homes and green gardens. Boys play on a softball diamond. The movie theater advertises a family-type show that any small-town American church group would endorse. Housewives shop in a well-stocked supermarket. Baby buggies line the walk before its entrance. The town is like a California desert community transplanted into the Middle East.

The Americanized town is quite a change from the oil camp that existed here in 1938. On one of my several visits to Dhahran, T.C.

"Tom" Barger still was president of Aramco. He is a wiry, athletic oil man who started as a geologist with Socal in 1937. He was immediately transferred to Arabia, arriving before oil had been found.

"And it looked as if we never would find any," he said. Ten dry holes had been drilled on the promising Dammam Dome near modern Dhahran without hitting oil. The Socal jinx seemed to be working again.

"The prospect of finding oil was so bleak that on arrival I was told that it was unlikely that I would be in Arabia for more than a year," said Barger. "We all expected to be shifted to Indonesia."

Max Steineke, the blunt-jawed geologist whose blown-up photo now stares down from a wall of the living room of Aramco's Dhahran Guest House, suggested that perhaps the wells had not been drilled deeply enough. It was decided to return to Dry Hole Number Seven and push further down. This was a fortunate decision.

The drill bit of No. 7 struck "large quantities of oil" at 4,727 feet in March 1938. Since this oil was only a few miles from the sea, Aramco was able to export the initial shipment in little more than a year after the discovery. That first shipment was exported on the tanker *D. G. Schofield* on May 1, 1939, with King Abd al-Aziz and most of his court on hand to join in the celebration.

Today, the well still is producing. Ahmed A. Lughod, a husky American of Palestinian descent, drove us to No. 7 at Barger's suggestion. A "Christmas Tree," an aluminum painted collection of valves and pipes, sat atop the well behind a wire fence. Oil gushes easily through those pipes and underground pressure is so strong that Aramco needs no pumping to lift the oil. It rises naturally at five thousand and more barrels per day.

Despite the international oil intrigues in the area, Aramco remained a 100 percent American operation until 25 percent equity participation was undertaken by the Saudi government on January 1, 1973.

Shortly after this initial discovery it became apparent to Socal and its Aramco subsidiary that they were sitting on one of the largest reserves in the world. Later exploration revealed that indeed Aramco holdings did include the largest-known petroleum reserves anywhere, holdings so great that before the end of 1974 Saudi Arabia could emerge as the world's number one petroleum producer. This likelihood depends upon whether or not Saudi Arabia meets America's need for oil, or limits production with the Arab world on the question of America's support for Israel. After the Yom Kippur War, Saudi Arabia didn't seem prone to help America in anything. Should Saudi Arabia forge to the Number One spot, the United States would drop to Number Two, ending a U.S. leadership that started with that first oil well at Titusville, Pennsylvania, in 1859.

But back in those pre-World War II days, it was evident that no one company could merchandise and sell all the oil that would be coming from Saudi Arabia. Socal sought partners, the normal thing in an industry in which capital expenditures are huge. That search brought it into collision with the Red Line Agreement.

Both Standard Oil of New Jersey (Exxon) and Socony-Vacuum Oil Company (Mobiloil) wanted to participate in the Saudi venture. Who wouldn't after analyzing prospects? They had markets that could be linked with the Arabian production, a contribution of importance in those days when an oil glut was appearing. However, IPC partners of Jersey Standard and Socony refused to allow them to violate the Red Line Agreement. Anglo-Persian, Shell, Compagnie Française des Pétroles, and Mr. Five Percent Gulbenkian did not want to help this upstart American company Socal, except perhaps by moving in themselves to share in the venture.

A tentative agreement was drawn in 1939 to do just that. Red Line companies would help sell Aramco's oil. This seemed to be a victory for Compagnie Française des Pétroles, for France had only Iraqi oil as an assured source at that time.

But World War II intervened before CFP and others could move into Saudi oil. With the war over, Americans contended that the Red Line Agreement was obsolete. Jersey Standard and Socony opened negotiations to participate directly in Aramco. The deal finally was closed in November 1948. Aramco's ownership was settled in its 1972 mold, Jersey Standard (Exxon), Standard Oil Company of California, and Texas Company (Texaco) each 30 percent and Socony (Mobiloil), 10 percent. On January 1, 1973, Saudi Arabia took its 25 percent. Companies' equities were scaled down according to percentages held prior to the government participation.

There were several distinct advantages that favored Middle East oil from the first. The autocratic governments made it possible for international oil companies to deal directly with rulers or their designees. There were no parliaments to raise pointed questions. Since oil and mineral rights belonged to the states along with most of the rural land, there was no problem with private claimants. It was possible for companies to establish huge concessions and to explore them en bloc. The Standard Oil Company of California's original concession in Saudi Arabia, for instance, covered an area of 495,000 square miles, or roughly the size of Arizona, New Mexico, and Texas combined.

Moreover, Middle Eastern oil is not too deep below the ground, about five thousand feet on average. It generally lies close to the sea, and pipeline and other costs are low. The reserves are enormous, and individual wells are prolific, around 6,000 barrels a day versus a United States average of 19 barrels daily.

On the company side, seven companies—Exxon, Texaco, Shell, Standard Oil of California, Gulf Oil, Mobiloil, and British Petroleum—managed to lock up most of the productive capacity early. This made it easy for cartel-like arrangements to be concluded through discussions among only a handful of people. By scratching each other's backs, these seven companies managed to hold 90 percent or more of the international oil business for years. Compagnie Française des Pétroles is a weak eighth among these giants, in league with the giants in several areas.

Nevertheless, Middle East oil production still was in its infancy as the 1930s drew to a close. British Petroleum Company statistics show that in 1938, the total Middle East output of 325,000 barrels of crude per day was as follows:

	1938 (In thousands)
Iran	210
Iraq	90
Bahrain	21
Egypt	3
Saudi Arabia	1

The oil search in Egypt had started as early as 1869, without much luck. Oil finally was struck at Gemsa, on the Gulf of Suez, in 1909. But Egypt has never been able to move into the front ranks of Middle East producers. The same can be said for Turkey. It was not until 1950 that the first oil was found in Anatolia. In 1973, domestic production was only meeting about half of domestic requirements.

It was after World War II, especially in the decades of the 1950s and 1960s, that Middle East oil output soared. In the latter decade Abu Dhabi and Oman followed Qatar among Persian Gulf sheikhdoms as oil producers.

In 1949, Middle East output in thousands of barrels daily amounted to the following:

	1949 (In thousands)
Iran	560
Saudi Arabia	480
Kuwait	250
Iraq	80
Egypt	40
Bahrain	28
Qatar	2
Total	1,440

By 1972, the Middle East had moved to the front of the stage. Its 1938 output had only been 5.5 percent of total world output. In 1972, the Middle East accounted for about 40 percent of the total, with that percentage showing all signs of soaring much higher in the years ahead. The output may triple in Saudi Arabia over the next decade, for instance.

The world energy shortage bolsters the power of producing nations. It is not only America that needs imported oil. Europe has always depended heavily upon Middle East oil. Japan has emerged as a great new industrial power and its consumption of oil is soaring. The Middle East provides the bulk of that.

Today, Middle East nations do not even need salesmanship to sell their petroleum. Their customers, the Americans, the Europeans and the Japanese rush to invest money in oil, then use their technology to lift it from the ground. Motorists at self-service stations only man pumps to get their gasoline; oil entrepreneurs not only man the pumps, they build or organize the entire paraphernalia of oil from tapping the subterranean reservoirs to refining and delivering the oil to customers.

Thus, the cash registers jingle merrily for Middle Easterners. They are being presented with soaring incomes that even Aladdin probably could not have conjured with his lamp. Nations of the Middle East earned $7.1 billion from their oil in 1971. That total jumped to over $11 billion in 1972 and to around $15 billion in 1973. In the period from 1972 to 1975 their earnings should exceed $80 billion. By 1980, Middle East nations may be earning as much as $60 billion a year from their oil.

Seldom in man's history has there been such a dramatic shift of income. Overnight developing have-not countries are joining the ranks of the haves. These nations cannot spend the money fast enough to transform themselves immediately into industrial nations. They do not have enough educated, trained people to fill all the necessary jobs. They are in the novel position of being *rich developing nations,* something that is most unusual in a world where the poor greatly outnumber the rich.

Soaring revenues for the Middle Easterner means skyrocketing prices for American, British, continental European, and Japanese consumers. The rising production in the Middle East could help meet the vital energy needs in the major industrialized countries. But will politically inspired supply interruptions proliferate?

It is a good question. When I asked it in Riyadh, Saudi Arabia's capital, I received the typical answer of Islam: *"Insha allah"* (If God wills). Al-Mutanabbi, an Abbasid dynasty poet, said, "In seeking honey, expect the stings of bees."

That has a bit of the cynicism found in Arab thinking. Often, it does seem as if you should expect the worst where the Middle East is concerned. Sometimes it seems as if there is an irrationality about actions that aggravates whatever problems are on the table.

It was Al-Mutanabbi, too, who wrote, "A drowning man cares not about getting wet." To understand the psychology of the Middle

Easterner, one must know that he does sometimes see himself as the drowning man even when surrounded by dry sand.

The Middle Easterner believes that oil has been underpriced for years. He thinks that a doubling of the price is only just, a belated correction of a distorted price situation. Once the higher price level is attained, he is apt to look backward in bitterness at the old price, using it to prove that he was exploited for years by the oil companies. He certainly will defend the higher prices even if it takes a disruption of the oil flow to do it.

Arab hatred of Israel also has been intensified by defeats in the Yom Kippur War of October, 1973. In that month, eight Arab oil producing countries were limiting shipments to America and other pro-Israeli nations. Israel, the real target, seemed unconcerned, and it seemed as if Arabs were striking out, wildly.

Irrational? It may sound so to the Western mind, but the Middle Easterner reaches conclusions through a combination of emotionalism tempered with Islam (or Judaism, in the case of Israelis). In the final analysis it will be the Middle Easterner who will decide how the West's need for petroleum is to be met. Unfortunately, there is no mass opinion about anything in this part of the world. Even such a sacred thing as a tenet of Islam may have several different interpretations, depending upon whether the interpreter happens to be a Sunni, Shiite, Druse, or a member of one of the Sunni divisions.

The Iranian, a non-Arab, is far different in many respects from the sons of Saudi Arabia or of Egypt. In other respects he has that same desire to right past wrongs, the long memory for slights, and the willingness to take punishment to attain an end.

VI

The Iranians

The turbaned beggar squatted on the shoulder of the Iranian high-way, where tawny plains were greening in the spring. Except for a peasant tilling a field with an ox team, only sparkling sunshine filled space between the beggar and the distant, snow-draped Elburz Mountains.

"How does he expect to collect anything there?" I asked, as our Plymouth whizzed by.

"It is a good place," said A. H. Jalali, an oil-company official. Over a period of days I had found that nothing disturbed him no matter what the provocation. Now he shrugged. "This highway is known as the 'Road of Death' because of its many fatal accidents. People think it is good luck to give a coin to a beggar. So many people will stop to give this fellow a few rials, hoping for good luck against accidents."

I remembered the two Volkswagens that had passed us a few minutes before, racing neck and neck as the overtaken driver refused to concede defeat. Fortunately, we had been crowding our side of the road and no oncoming traffic appeared.

"A surer way to avoid accidents would be to drive more carefully," I said.

"Ah . . . that is the logical approach," replied Jalali. "But many of our people are not logical."

He was so right. The typical Iranian values pride more than logic, his idea of honor more than untroubled peace. He is likely to keep his word once an agreement has been made, but he relishes bargaining so much that almost every purchase through a day may be made only after bitter haggling over the price.

In 1951, Iran nationalized its oil industry in an unplanned gesture that smacked more of pique than forethought. Illogical perhaps, yet it also reflected the intense pride and strong nationalism that is part of every Iranian. This country's history goes back 2,500 years, and Iranians do not intend to forget that, especially where foreign exploitation of their oil is concerned.

Nearly all Middle Easterners have a comparable pride, even if known antecedents do not extend as far into antiquity as those of the Iranians. The Iranians, having been independent longer than other countries in this part of the world have had more opportunity to display that pride.

While nationalization was irrational to the Western mind, it was not to the Iranian. This is a nation struggling to assert that independence after centuries of political unrest and economic stagnation. When Premier Mohammed Mossadegh nationalized the oil industry in 1951, few Iranians worried about the cost to the nation or about the loss of production stemming from counteraction by international oil companies. Mossadegh had made a popular, crowd-pleasing gesture. Street mobs cheered.

Iran is a nation of over thirty million people with an area of 630,000 square miles. This is equivalent to the total area of Oregon, California, Nevada, Utah, Arizona, and Washington. Its twenty-five centuries of recorded history started with the Achaemenid Empire (550–330 B.C.), which welded about twenty-seven nations together to form an empire stretching from India to Greece. It was Cyrus the Great who founded this dynasty, the same Cyrus who is mentioned favorably in the first chapter of the Book of Ezra and the Book of Esther. In conquering Babylonia, Cyrus freed the Jews who were being held in captivity, allowing them to return to Jerusalem. In an age of violence, he believed in religious tolerance.

In appearance, this land is much like the American Southwest, with flat stretches of dry plains broken by rolling hills and mountains. But no mountains in the American Southwest soar like those in Iran. Mount Damavand, an extinct volcano, rises majestically behind Tehran to a height of 18,934 feet, and falls away on the north side to the Caspian Sea, which is eighty-five feet below sea level. On this coast, Iran resembles a South Seas landscape. Thatch huts are buried amid banana groves. Bamboo thickets shield quiet pools. Tea bushes form orderly green rows on hillsides. Women in colorful skirts pluck the tender leaves, each working with a straw basket on her arm.

Iranians call the lush country beside the Caspian "the jangal," which indicates where we get our word "jungle." "Bad" is spelled the same in English as it is in Farsi, the language of Iran. A bank is a "bânk." "Sigar" is a cigaret. The linguist finds other similarities when he pairs root words, for Farsi comes from the same Indo-European base as do most of the languages of Western Europe. The Persian, or Iranian if you will, himself, comes from the same Caucasian stock as the Angle and the Saxon. The name "Iran," adopted in 1935, means "the land of the Aryans."

"We are not Arab," I was told again and again by Iranians on several visits to the country. The average Iranian resents being mistaken for an Arab, for no love is lost between the two peoples. Iran's closest neighbors, the Iraqis, certainly fan this feeling. Both nations claim stretches of the Shatt-al-Arab, the river that forms the boundary near Abadan. Border disputes elsewhere occasionally result in an exchange of gunfire. Iran claimed Bahrain for years, though it recently renounced its suit. When a dispute arose concerning the ownership of a few tiny islands, possibly containing oil, in the Persian Gulf between the United Arab Emirates and Iran, the latter country occupied them.

Iran further aggravates Arab sensibilities by maintaining good relations with Israel. It even ships some of its oil to market via pipeline from Eilat on the Gulf of Aqaba across Israel to an outlet on the Mediterranean.

The Shah's attitude toward Iraqis is clear-cut. Says he: "Just imagine a country of savages putting an end to an old people like us with all our future before us." The Shah is not a man to equivocate. Now that the British have withdrawn military forces from the Persian Gulf, Iran is the dominating power in this part of the world. The Shah intends to keep it this way, and he has the support of his people behind him.

Despite Iran's military buildup, modern Iranians are not militaristic. The country has the strongest army in the region if one does not look north to the Soviet Union, but this army spends most of its time building roads to isolated villages, repairing bridges, and acting as policemen in lonely areas. There even is a Literacy Corps, about twenty thousand draftees who spend military service teaching in village schools about the country.

This is all part of Shah Pahlevi's "White Revolution," a program that is revolutionizing the country through economic development.

At fifty-five, Shah Pahlevi is a handsome, brown-eyed ruler with an athletic build who evolved from an international playboy into a dedicated social reformer during the thirty-three years he has been in power. His shock of once-black hair and sideburns are streaked with gray. He obviously needs glasses to recognize anyone he faces in an audience. Still he dons his black-rimmed spectacles almost with reluctance, as if unwilling to highlight any of his weaknesses. A licensed pilot, a crack skier, and a good football player in his youth, he is a man of action who believes in attacking a problem head on.

His office attire is a nattily tailored charcoal suit, probably with a plain royal purple tie. His manner is urbane. He seems eager to put a commoner at his ease. Yet he has the quiet authority of a man born to command, then trained all his life to assure his position. The Shahanshah (king of kings) of Iran is *the boss* of his country in title and in fact, a man who feels that he has been destined to bring his country from the medieval age into the atomic era. And that destiny has been established by Allah.

Oil, his country's greatest natural resource, means money for the economic development of his country. His voice grows impassioned when he talks of the need for obtaining about $36 billion from oil in the 1973–78 period to finance dams, agricultural projects, roads, schools, factories, and other developments within his realm. Thus it is almost a personal crusade for the shah to squeeze more and more money from foreign oil companies. He is critical of international companies that have high production in low-population nations when a country with Iran's large population vitally needs more revenue. Thus, oil revenues go to little sheikhdoms which can't spend the money whereas Iran has substantial oil reserves to exploit.

The Shah's White Revolution already has achieved spectacular results. On the campus of the University of Tehran, a tousle-haired student with wisp of moustache proudly showed me through a cluster of modern buildings aswarm with thousands of students. "This is where the new generation of Iran's leaders is coming from," he said. He himself was hoping to obtain a position with the National Iranian Oil Company on graduation. "There are too many foreigners working in administrative positions in our oil industry today," he said. "It is time we Iranians took over these jobs."

I got the impression that he probably wanted to start at the top, just as do many youngsters fresh from college in the United States.

In a rocky canyon on the mountain road between Tehran and the Caspian Sea, a huge dam blocks the gorge. A guide enthusiastically cited statistics of power and irrigation for the project, one of many rejuvenating the country.

In an office in Tehran, an economist presented more statistics to show that the country's gross national product is rising at an 11 percent annual rate. This is one of the highest in the world. He was smoking a water pipe as he shuffled the papers on his desk, the water container sitting on the carpet beside his desk.

Nobody can show more enthusiasm for the White Revolution than can the Shah himself. He sometimes seems surprised that international oil companies do not support more willingly his ideas for sharing the oil wealth with the people of his country. There is an edge in his voice when oil firms enter the conversation.

In one interview, he bluntly called for the elimination of the middlemen between producer and consuming countries in international oil. This, of course, is the role played by the international oil companies. "The question is to find a way to arrive at that without creating too much difficulty at the beginning," he said. "Obviously, this cannot be done overnight, but maybe we should think about this."

Oil companies were thinking about it too in early 1973 as their negotiations with Iran hung fire. When Iran ended by taking over properties of oil companies in the summer of 1973, company officials could reflect that Iranians are very tough bargainers, whether the deal involves a million barrels of oil or a half dozen eggs from a corner merchant.

After a particularly rough bargaining session involving a loan for Iran, one World Bank official emerged from an office in Tehran,

exhausted. "The Persians drive harder bargains on credit than anyone I have ever dealt with," he said.

It was all right for him to refer to Iran as Persia. The present shah
has decreed that either is correct, though Iran now is preferred by
most people. In one World War II encounter with the Shah, Winston Churchill grumpily said: "No one will ever intimidate me into
speaking of Persia in any way except Persia." His eyes twinkled,
however, when he said it.

But whether you call them Persians or Iranians, they are equally
good at bargaining. Mullah Nasreddin, a joke figure in Iranian lore
akin to the Pat and Mike of Irish storytelling, illustrates the Iranian
knack for seeking the better side of a deal.

One day Mullah Nasreddin asked a music teacher the charges for
taking lessons. "Three silver pieces for the first month," the music
teacher said. "One silver piece a month thereafter."

"In that case," said Mullah Nasreddin, "I would like to start with
the second month."

Centuries of invasions have left many minorities in the country.
There are Kurds, Armenians, Indians, Afghans, Arabs, and various
other groups. Often minorities live in their own communities, speaking their own languages, following their own customs.

"We are a tolerant people," I was told in Isfahan when I commented about the many diverse peoples in Iran. In that city, we
encountered a Jewish colony, which legend says has been there since
the time of Nebuchadnezzar, about 700 B.C. There are an estimated
six thousand of these Jews living together in the Joobareh quarter
of the city, undisturbed by their neighbors. "We have thirteen
synagogues," one bearded Jewish elder said through an interpreter.
The colony is declining, however, through migrations to Israel.

About twenty-one thousand Zoroastrians still practice their fire-
based religion. At the Atash-i-Varahran Fire Temple in Yazd, a
white-robed priest tended the sacred fire which he insisted had been
burning constantly since long before time of Christ. He explained
that in ancient times there were four especially sacred temples in the
country. These were located over places where natural-gas seepage
provided a perpetual fuel for the sacred fires.

With his white smock, gauze facial mask and white skullcap, he
looked more like a surgeon about to stride into an operating room

than a priest of an ancient religion. Although this religion is sometimes described as fire worship, it is not. Fire and light represent good and holiness to the Zoroastrian. Fire is a symbol of good rather than an object of worship in itself. The creed of this religion is summarized in this simple precept: "Think good, speak good, do good."

Not only are Iranians tolerant about other religions, their own Shiite Moslemism is tolerant too. But there are certain shrines in the country—such as the pilgrimage city of Mashhad, far to the east of Tehran—which are best left unvisited if one is a non-Moslem. This is the holy city of the Shiite religion, a place as venerated as Mecca. Its shrines are taboo to the non-Moslem.

The Shiite believes that Ali, the son-in-law of Mohammed, inherited the caliphate, or leadership of Islam, when Mohammed died. The Sunni believes that this position cannot be inherited but that the caliph must be selected by Islamic leaders. From original differences, branches of Islam diverged. Today, Persia and Iraq are centers of Shiite practices.

Generally, one encounters friendliness nearly everywhere. Tea invitations are numerous, in shops, in villages where one is a stranger, or when pausing beside a nomad's tent on a rolling plain. It is the custom to accept, too, even when one is in a hurry. For, according to the Iranian view, one should never be in so much of a hurry that one cannot pause to make a new friend. If the business at the other end is delayed a little while, what matter? Shoes are removed. One enters the goat-hair tent, the village hut, or the town house, and relaxes on the carpet or on a cushion. Tea arrives in good time, with lumps of sugar and a little lemon.

Iranians certainly are more tolerant of alcohol than are most Moslems. One night in the Sha-er Grill of the Hilton Hotel in Tehran, we were introduced to 1001 wine by three Iranian friends. They insisted that this was the wine of the *Thousand and One Nights* of Arabian mythology. The evening was well along when I asked for a translation of the Arabic inscription carved in the polished brass over the restaurant's rotisserie.

Roughly the translation read, "We are men of wine, food and music. Happiness is ours when there is wine, food and music." Omar

Khayyám, the poet and mathematician, caught this spirit in his Rubâyat:

Here with a loaf of bread beneath the bough
A flask of wine, a book of verse, and thou
Beside me singing in the wilderness—

Khayyám, a tentmaker's son, was born in the twelfth century A.D. at Nishapur, spending his life in that city. Curiously, he is honored more for his mathematical genius in Iran than for his poetry about the illusory nature of life.

Ah, make the most of what we yet may spend,
Before we too into the dust descend.

A love of poetry is found at all levels in Iran. The simple peasant in one of the country's sixty thousand villages may not be able to read. But he will appreciate the storyteller who sits in a coffee shop reciting poetry for a few coins from his audience at the end of his recitation.

Once in a small village high in the mountains behind Tehran, I stopped at a country school on a knoll beside a row of poplars above a swift-running stream. Here I visited one of the four classrooms in the mudbrick building. The schoolmaster, an elderly gentleman in a frayed western suit, met us on the steps, beaming hospitably. Behind him two dozen boys with shaved heads stared at us, round-eyed, slate boards in hands.

He had been reading selections from Firdusi's Book of Kings to the boys, an epic tale of Rustem, the mighty warrior. The professor spoke a little English but was afraid to use it.

"In English, I am stupid," he said. So we conversed through the Ministry of Information interpreter with me. "Firdusi is such a great poet that it gives me pleasure when I can read his verses to my class," said he. "Tell me, what do people in America say about his works?"

He had so much enthusiasm for the poet that I could not tell him that Firdusi is almost unknown in America. "Everyone who reads him likes him," I said lamely. He beamed, satisfied with the answer.

On a drive from the Caspian Sea over the mountains to Tehran, the driver suggested a rest stop. He made the suggestion by citing an old Persian saying, "A halt on the journey, a rest, a drink from the well, and the caravan moves on at the setting of the stars."

Shiraz, the minaret-studded city in the south, is the cradle city of Persian poetry. This "city of roses and nightingales" was the home of two of the country's greatest poets, Hafiz and Sadi. Hafiz, a fourteenth-century poet, was the lyrical Keats and Shelley of this country's poetry, a man who wrote with feeling no matter what his topic.

Sadi is no less popular, not only in Iran but throughout the Arab world too. He lived for a century from about A.D. 1181 to 1291, and mixed philosophy with his poetry, leaving behind a volume of work that is quoted daily by storytellers, men of letters, and beggars.

The Iranian is artistic by nature, too. The world has known no finer silk, velvet, carpets, textiles, miniatures, tilework, pottery, mosaic work, and embossed silver than that which may be found in this country's art shops. And Isfahan always has been this country's artistic heartland, the place where literally thousands of hand crafts-men and artists labor for days and weeks with their brushes and hand tools to produce that ivory-inlay jewel box, that miniature painting of a medieval scene, or the delicate carpet that owes its sheen to the slightly off-center knots of the hand-tied fibers.

In Tehran, at the Honorhaye Melli, the government institution that is fostering development of native arts, I once spent a day photographing the work and talking to the handicraft artists. Here, in a palace of the Qajar dynasty, one can see examples of rugs, carvings, brocades, ceramics, paintings, and other art works.

I watched an ivory carver bending over his bench, fashioning a delicate picture frame that I was told would take months to com-plete. This is the sort of task for which the western worker no longer has any patience. Skillfully, the carver worked, shaving a hair-thin piece here, gouging a pin prick there. He took pride in his actions, especially when he saw me focusing a 35-millimeter camera with some 800 A.S.A. film for some unobtrusive shots.

"After spending months on such a lovely piece of work, it must be difficult to give it up," I said to him.

He smiled and said, "When I finish something, I take one last long look at it. That is compensation enough."

Time is rated differently in Iran than it is in western workshops.

Even the calendar is different. Iran dates its calendar from the Hegira of A.D. 622, as the Arabs do. But the Iranian calendar is adapted to sun time with the New Year beginning on March 21. Thus the Iranian year 1352 A.H. started on March 21, 1973.

Iranians launch their year with the thirteen-days *Now Ruz* holiday, a period when nearly everything stops. Offices are depopulated, workers call in sick, and shops may be closed most of the time. It is a repetition of the Moslem Ramadan without the religious overtones.

Now Ruz reaches its climax on that thirteenth day, the day of *Sizdah*. This is a day unlike anything to be found anywhere except perhaps in the *Hajj* at Mecca. It is a picnic day when everybody feels he must leave home for the day, regardless of where he lives. And he takes the whole family with him.

The belief underlying *Sizdah* is that the thirteenth day of this first month is unlucky. The safest and wisest thing to do is to avoid this bad luck by being away from home all through the day. Everyone, regardless of job, social rank, or locality, tries to be in the countryside picnicking. Those with automobiles offer rides to friends and relatives. Every bus and truck is drafted by charter groups or by commercial travelers. Carts, mules, donkeys, bicycles, and everything else provide additional transport.

The exodus looks like the panic reaction to a great natural catastrophe or an invasion. Lines of motor vehicles extend for miles on roads, and vehicles scatter over the countryside. Soon every tree has its contingent of picnickers.

By tradition, wheat or lentil seeds are planted in a bowl at home several days before the day of *Sizdah*. The tiny sprouts are uprooted. Each member of the family is supposed to cast one sprout into a stream somewhere in the country. This, according to legend, will carry away troubles.

Anyone unfortunate enough to be tied to his city on the day of *Sizdah* spends uneasy hours, haunted by fear of unforeseen catastrophe. Only at sunset are those fears eased. By that time, in a city like Tehran, most of the population will be on a bumper-to-bumper highway struggling to return to their homes. Many do not make it until long after midnight.

Undoubtedly, the roots of *Sizdah* extend far back in time. Perhaps in man's nomadic past this was the day when the winter encampment was broken with the spring trek to new pasture. It might have been "bad luck" to remain longer at the winter camp, as spring rains carried the camp's disease-bearing filth into the water supply.

In Tehran, a city of three million, it is easy to forget that Iran is basically an agricultural country. The bulk of the population lives in the country's sixty thousand villages. Such a community is Barquijan, set in a valley of the Elburz Mountains, thirty-five miles from Tehran. Its population is about fifteen hundred. Its way of life is akin to that of A.D. 1500.

You reach Barquijan bouncing over miles of a rocky road that discourages speed. Barren hills roll upward to snow-covered peaks nearly nineteen thousand feet high. Swift streams cut through green valley floors. The road twists and turns, mounts and descends.

On the edge of the village, the wind rippled a field of green barley, creating wild patterns of motion. Inquisitive children dashed forward, shouting. Girls wore pajama-like pants beneath bright dresses. Boys had *shalwars*, those baggy pants that seem designed for fat men. Skullcaps sat atop close-shaven heads.

Ebrahim Toosi, the village *kadkhuda* or headman, emerged from the cluster of mudbrick houses where the road became a rutted, unpaved street. He is a slender, black-moustached man who looks as if he might shave once a week without being aware that his schedule affects his appearance between shaves. There was a quiet dignity in his manner, though his unpressed western suit had the shapeless form of a blanket draping a pole.

"Welcome," he said to us through the Information Ministry official, who also was our interpreter, guide, and lecturer on the history and sociology of the country. "Allah has been good to us. You must see our school."

Villagers crowded around as he led the way on an inspection tour of this community, which is a microcosm of rural life in Iran. Goats grazed by a brook that twisted around the village. Tall poplars towered in a meadow near the one-story brick school. Steel beams for the roof had been carried over the mountain on shoulders of the villagers. Four teachers provided by the government of his Imperial

Majesty Mohammed Reza Pahlevi, shahanshah of Iran, taught about 125 youngsters in six classrooms.

At present about 90 percent of school-age children in the cities and 43 percent in rural areas are studying in primary schools. Educational spending plays a major role in the development spending of Iran's oil money. The need for development of all kinds provides the incentive for Iran to seek more and more money from the oil companies.

Only in comparatively recent times has the peasant family seen a need for educating girls. Now you find young girls toting books on their way to school. In Tehran girls are beginning to occupy many office jobs, and there is a feeling that women are being emancipated, but the militant female is nonexistent in this man's country.

"The god of women is a man. Therefore, all women must obey man," says one Iranian proverb. Another conveys the same impression: "Woman is a calamity, but no house ought to be without this evil."

In villages, October is the marrying month. One engaged youth in Barquijan explained, "The snows come in late October, and then we have four months to get acquainted."

He already had presented his thirteen-year-old bride-to-be with a mirror, the Iranian engagement symbol. The mirror will always be the girl's possession, the only household effect a wife may call her own in this Moslem land. As is usually the case even in towns, his bride was selected for him at a family council. This did not seem surprising either to him or to his girl friend, a shy, long-legged girl who did not join in the discussion.

Barquijan is a village of small landowners. Villagers have that independence of people who own their own land. Prior to the present regime, most of the agricultural land was in hands of landlords, many of them absentee. It used to be said that a thousand families controlled the whole country.

Shah Pahlevi instituted a land-reform program several years ago. He started by giving away his own private holdings. Then he energetically promoted a national agrarian reform, which gave thirty million acres to landless peasants, considerably reducing the power of landlords. The "Thousand Families" still are powerful in Iran, but there now are thousands of people who have moved into this country's elite through a spreading of opportunities. Moreover, the annual per capita income of $600 early in 1973 is expected to double in real terms in another five years. Whether or not opponents of the Shah care to admit it, he is spending his country's oil money wisely for benefit of the people. The Shah's strongest opposition comes from the far right, which sees its power being clipped, and the far left, which ridicules reforms as a smokescreen to perpetuate the monarchy.

The dedicated Communist is not interested in seeing someone else help the underprivileged. He wants all the credit for himself if there is any. Thus social welfare from the right of center must be ridiculed and even opposed.

Undoubtedly, conditions still are primitive in many of the thousands of villages in this country. Barquijan was no different in this respect.

Until ten years ago there was not even a road to this village. Everything had to be brought in or out via donkey trail. Even today a bicycle is a luxury, and the village's economy is largely subsistence. A few days in the village is like a visit into the past. Start the day in a mudbrick hut and you find yourself awakened at 4:30 A.M. while lying on a thin mattress spread on a Persian carpet.

Breakfast in the cheery glow of a kerosene lamp consisted of cheese, sangak (unleavened bread), and warm goat milk. The heads of the children clustered around the table made large shadows on the walls of the hut, where spare clothes were hung on pegs driven into the wall. The hand-made table and chairs were crude but serviceable. A colorful curtain served as the door to block off a bedroom where six youngsters slept each night on the floor.

In the early morning light, two boys milked several goats in the compound before the house. They were nearly finished when villagers trekked to the fields. Though land is worked on an individual basis, villagers often share the work, helping each other. An irrigation ditch was cleared of mud. Water was diverted into a channel. Weeds were hoed in cabbage and melon patches. Boys drove goats to pasture, then raced to a cherry orchard to pick the ripe fruit. School was for later after the fruit was in.

Lunch was at noon, the biggest meal of the day, shishkebab with rice and tea. Then came a welcome two hours' nap in the heat of the day. In this mountain valley that heat was more illusory than real. Work resumed after the siesta to continue until dark. Goats and cows again were milked before the evening meal of bread, cheese, and meat stew. By that time, even a mattress spread on the hard floor was inviting.

"We have a good life," said Ali Toosi. He was an amiable forty-year-older who seemed satisfied with little, unenvious of others, ready to accept what Allah provided. He had a crop of thick black hair that merged into a week's growth of beard. As the janitor at the school he earned fifty dollars a month from the government. This made him the highest-paid man in the village, and he was willing to discuss his good fortune over lunch one day in his home.

It was a cool, heavily carpeted room. Several cushions against a wall were the only furniture. Guests sat on these, making sure that feet were tucked crosswise. It is considered rude in the Middle East to sit with legs distended, soles of the feet facing the host. In villages and in old-style homes in Iran, shoes are removed at the door. Inside, you walk barefoot or in stocking feet.

To an Iranian a rug is a status symbol rather like the American automobile. A man buys as many carpets as he can afford. An Iranian with an earthen floor is at the bottom of the social scale. Even a nomad in a goat-hair tent usually can afford at least one carpet.

There are five basic weaving designs in Persian carpets: hunting, garden, vase, prayer, and medallion. Contrary to what you might think, those carpet designs are not a haphazard collection of geometrical designs in colors. They are symbols, each having a special meaning to the initiated. Look closely and you may see a tree interwoven in fibers of a certain carpet. It is the tree of life, wishing long life to the possessor of that particular carpet. A triangle is considered lucky and, hopefully, some of that luck in a carpet will apply to the owner. A pomegranate design is aimed at increasing the fertility of the family.

Colors have their own significance. Green is a sacred color, Mohammed's favorite. Red signifies life and happiness. Blue is the warrior's color, signifying power and might. Black is for sorrow. And white personifies goodness.

Age adds value and wear gives more "life" to a carpet than does a rest in a warehouse. In Tehran you sometimes see carpets spread on pavements in the middle of streets. Passing traffic "ages" them in a few hours, perhaps doubling values in subsequent bargaining sessions with foreigners who may be seeking old Persian carpets.

At least 10 percent of Iran's population still live as nomads, depending upon flocks of sheep and goats for livelihood. Another substantial percentage may take to tent life for summer pasturing of flocks in the mountains. Tribes include the Khamseh, the Kashgai, the Lurs, the Kuh Galu, and others.

Near Persepolis we encountered a Kashgai clan on a trek from winter pastures to the mountains one March day, during one of numerous trips to that part of the world. Hundreds of bleating sheep and goats raised a great cloud of dust. Men in blue robes, their felt-hat ear flaps hanging down, herded the animals across the arid plain. Their ponies snorted and pranced, harnesses jangling, a rider occasionally making a path between white bodies of sheep.

Ragged youngsters and women in long, multicolored dresses ran along behind or rode sidesaddle on aged horses no longer considered fit for men to ride. A string of loaded camels carried the tents and camp gear of the group.

Late that afternoon, we visited their encampment beneath the cliffside imperial tombs of Naghsh-E-Rustam. Camp fires flickered in the late afternoon light, women clustered beside simmering pots, none veiled. Kashgai women have much more freedom than do the usual rural females. Now some of them were busily straightening carpets on floors of low goat-hair tents that had already been erected. The clan chief rose to greet us, extending a bony hand.

He was the color of saddle leather, his grizzled features wrinkled from the sun rather than from age. Perhaps about forty, he had a muscular grace and a youthful vigor that indicated he probably was a lot younger then he appeared.

He apologized because he could not offer us tea. A pack horse containing the tea had bolted and several boys now were searching for it. But he had buttermilk. Would we join him?

When encountering the nomad in Saudi Arabia, in Iraq, in Syria, in Iran, and elsewhere, I have always been intrigued by the question: Is the nomad a wanderer by choice, or because of his pinched economic circumstances? I have received different answers at different times.

"He is a wanderer because he has no home of his own," one agricultural official insisted in Riyadh. "Give him a place to stay and he will settle down."

"Not so," said one date farmer near Hofuf in that big al-Hasa oasis in Saudi Arabia's Eastern Province, where man had settled three thousand years before Christ. "A Bedouin is a nomad by nature, like a gazelle in the desert."

I asked that question of this Kashgai chieftain as we sat cross-legged on the tent carpet. A trio of his aides sat beside him, eyes fixed steadily upon me. I wondered if the chief would understand my question. Does a tribal chief spend any time thinking of the meaning of life, and about the why of things?

I should not have wondered. His eyes kindled at the question. "Does the stallion welcome the bridle?" he asked. "Does the leopard seek the cage?"

Two of his sons worked at the Abadan Refinery. He realized that the old ways were changing, that the government was settling nomads on farm land, that perhaps his sons or grandsons would not like the life of his forefathers.

"For me, I will settle down when the wind stops blowing," said he. "Farming is women's work."

Nomads, however, do not roam like the gypsy. Each tribe has its definite area, whether in Iraq, Syria, Saudi Arabia, or Iran. Each clan and each family knows its pasturage in the tribal area.

"I can look at the calendar and tell within a mile where my family will be at any time of the year," Isa Qasqai, a student at Tehran University said. He had insisted that we visit his nomadic family near Isfahan. That had brought us to this encampment.

Before going to the camp, I had been told not to express any admiration for any of the tribe's possessions. Etiquette here and elsewhere in Iran calls for the host to present the guest with any article that is much admired. This is a social custom not always maintained, it is true, but a host may feel ashamed of his own poverty or miserliness should he fail to observe this custom.

Iran has had a constitutional government since 1906. But its

Majlis, or assembly, has not been a good example of democratic government in those times when it was free to express itself. In 1921, Reza Khan, the present shah's father and an army officer who worked his way up through the ranks, staged a coup to seize power. In 1925 he installed himself as emperor, ending the Qajar dynasty.

Reza Shah, an energetic, hard-driving man, ruled a country where each tribal chieftain was a law unto himself. He restored a semblance of order and strengthened the power of the central government. He had his troubles with Anglo-Persian Oil Company, too. On November 27, 1932, he cancelled the oil concession. Britain took the case to the League of Nations, but before the League could make up its mind about what to do, a new agreement was negotiated between Persia and the company. Shortly after, in 1935, Persia changed its name to Iran. Anglo-Persian followed suit, changing its name to Anglo-Iranian Oil Company.

During World War II, Britain and the Soviet Union accused Reza Shah of pro-German activity. They occupied the country, forcing the abdication of Reza Shah in 1941 in favor of his son, Mohammed Reza Shah, the present emperor. Foreign troops were withdrawn in 1946 after a short-lived, Soviet-supported "autonomous" movement in Iran's Azerbaijan province. This movement was condemned by the United Nations, and the Soviets withdrew.

But the U.S.S.R. continued subversive activities, channeling funds and propaganda behind the left-wing Tudeh Party. After Iran denied an oil concession to the U.S.S.R. in the late 1940s, the Tudeh Party, parrotlike, launched a campaign against British exploitation of Iranian oil. This was a shrewd move. People generally resented Anglo-Iranian, and it was easy to stage demonstrations against the company. Newspapers called for nationalization, and the stage was set for Mohammed Mossadegh, a wily, left-wing politician.

On March 16, 1951, the Majlis voted to nationalize the country's oil. In the subsequent squabbling, the government fell, and Mossadegh swept to power as premier. He aimed at both toppling the Shah and ending foreign control of Iran's oil. Nationalization was used to fan popular support for both moves and he almost succeeded.

The Shah, then, lacked real power. Between the "Thousand Families" and the *Majlis*, he sometimes seemed to be little more than a figurehead caught in the spidery web of Middle East intrigues, and narrowly escaped one assassination attempt. Mossadegh seemed to have all lines of power as nationalization became a fact in May 1951.

But nationalization had been an unplanned crowd-pleasing gesture. Iran did not have the manpower to operate its oil industry, the markets to sell it, or the tankers for delivery. So the industry shut down and other producing countries took Iran's markets.

Britain severed relations with Iran, and nationalization became a major international diplomatic hassle that dragged on and on. Meanwhile Iran suffered. Government revenues plummeted, lack of imports hurt dozens of industries, unemployment mounted, and popular unrest spread.

Events reached a peak in August 1953 when Mossadegh promoted a rigged election that would have put all power in his hands. A popular uprising swept Mossadegh from power, and with the army and popular support behind him, the Shah took control and has remained in power since then.

Typical Iranian bargaining ingenuity soon ended the oil impasse. Though the National Iranian Oil Company would own the oil, a consortium of international oil companies would operate production and refining facilities. Oil would be produced on behalf of NIOC, which then would sell it to individual members of the Iranian Consortium. To break the British stranglehold, American companies and the Dutch-British Shell Group would join Anglo-Iranian to form the consortium.

The contract was signed on October 29, 1954, and was to expire in 1979. One month later Iran's oil industry was back in production. Anglo-Iranian changed its name to British Petroleum Company, Ltd. It received a compensation of £25 million sterling from Iran for lost assets, 40 percent of the new consortium equity, $32.4 million from its new partners, and an overriding royalty established to assure compensation of $510 million.

Consortium member companies and their interests are the following:

British Petroleum Company Ltd.	40%
Shell Petroleum N.V.	14%
Gulf Oil Corporation	7%

Standard Oil Company of California	7%
Exxon Corporation	7%
Texaco, Inc.	7%
Mobil Oil Corporation	7%
Compagnie Française des Pétroles	6%
The Iricon Group	5%

Originally there were sixteen smaller American companies in the Iricon Group. This was increased to seventeen in 1955. Then, various companies merged over the years. Six companies were in the group at start of 1973:

American Independent Oil Company
Atlantic Richfield Company
Charter Oil Company
Continental Oil Company
Getty Oil Company
Standard Oil Company (Ohio)

In his book, *Mission for My Country*, (Hutchinson) the Shah wrote that Mossadegh's fatal mistake lay in his "stubborn insistence that he knew how to market our oil with no help from foreigners. Yet at that time we possessed not a single tanker, nor did we have even the beginnings of an international marketing organization."

That lesson sank home not only in Iran but in every other oil-producing nation as well. Producing nations established the Organization of Petroleum Exporting Countries (OPEC) in 1960 to provide strength through unity. They insist companies must train nationals, and they have established marketing companies, tanker fleets, and other facilities for handling their petroleum.

Education became an obsession with the Shah in his plans for developing the country. No nation can control its resources unless it has the trained manpower for it. A developing nation often complains it is being exploited by foreign companies. It may merely be suffering from its own deficiencies, paying the price to foreigners for having them corrected. Correction may come after revenues obtained from foreign companies are poured into the infrastructure and the educational system of the host country.

The Shah's partners in OPEC are mainly Arab, and Iran's relations with Arab nations, such as Iraq, are none too good. But in OPEC, Iran takes a coldly commercial view of things. Politics are not allowed to intrude where oil is concerned. The main aim is to squeeze more money from the consuming world for its petroleum.

Iran's disdain for joining Arab causes means that it is highly unlikely that it ever would join in any broad politically inspired boycott against America. Yet oil production and demand are so finely balanced that any output disruptions at all of output might be disastrous. Moreover, the jealousy between Iran and Arab nations leads to leapfrogging of oil agreements. If one side thinks the other has a better agreement, the aggrieved party wants to tear up its agreement with companies and renegotiate on a higher level. Iran does not want to permit Arabs to claim more revenue than it is receiving, and vice versa.

In 1963, the Shah launched his "White Revolution." Its money requirements prompted him to take the lead with other OPEC nations in squeezing ever more lucrative agreements from companies. The bargaining reached its climax in the Tehran Agreement of February, 14, 1971, so called because negotiations took place in the Iranian capital. It provided for a 33 percent rise in posted prices of Gulf oil plus a guarantee that prices would be raised by five cents plus 2.5 percent on June 1, 1971, and by the same amounts on January 1, 1973, 1974, and 1975.

The agreement added $3 billion to company costs in 1971 plus an estimated additional $2 billion a year until 1975. For oil countries it meant a 50 percent increase in revenues. Iran's oil income amounted to $1.8 billion in 1971, soared to $2.7 billion in 1972 and to $3.3 billion in 1973. The figure will exceed $5 billion in 1975.

All oil-producing nations won benefits in new contracts scaled along Tehran lines. This still did not satisfy Arab oil producers in the Gulf, who wanted at least 51 percent ownership of oil companies within territories to provide them with control. An agreement negotiated in late 1972 provided for participation of countries to start at 25 percent, rising in steps to 51 percent in 1982.

In January 1973, Shah Pahlevi reopened negotiations with the Iranian Consortium, seeking a contract as good or better than that of Arab lands. Under this deal, obtained after hard bargaining with companies, Iran took over all oil properties of the consortium in the country, including the giant Abadan Refinery, an action ratified by the *Majlis* August 3, 1973. In return, companies got a 20 years guarantee of oil supplies in proportion to their interests in the consortium.

The Shah wanted the added revenue to finance a grandiose new 1973–78 five years development program. This called for an annual growth rate of 11.4 percent a year, one of the highest, if not the highest in the world.

Critics of the Shah are quick to point out that, for all his interest in improving the lot of his subjects, he is an autocratic ruler who muzzles the press, crushes political opposition and discourages criticism. Drug peddlers are shot. Political activists may be.

Critics haven't suggested anyone better for Iran than the Shah. Even if they did it is doubtful that the Shah would listen. Unless overthrown by some revolution, which seemed remote in 1974, the Shah probably will continue as his nation's benevolent dictator as long as he lives.

After the Shah, what?

That is a question that oil companies ask. The Shah has been demanding on companies, he has squeezed harder and harder for cash, yet he has maintained stability in his country. Moreover, he is a factor for stability in the whole Gulf area—a bold, imaginative man who is not afraid to take chances for what he believes to be right. He is not an Arab lover, yet he has helped the Sultan of Oman in the latter's battle against far-left Marxist rebels in Oman's state of Dhofar. If Iraq should attempt to annex Kuwait, as Iraq sometimes has threatened, Iran might intervene to keep the little sheikhdom from being absorbed by its neighbor.

He already has ruled for thirty-three years, and he is a mortal man. Lovers of democracy may not like the Shah. Iran has never really known democracy, nor has any other country in the Middle East, except for Lebanon and Israel. So the Shah's passing could lead to chaos in the area, one more worry for the United States insofar as its energy crisis is concerned.

Strong men may rule well, but it is in their passing that one sees defects of their system when measured against democracy. Unfortunately, democracy, except for that of the Bedouin tent and of the family council, is unknown in this part of the world.

True democracy can only come when the opposition is willing to accept defeat within the framework of a nation's constitution, and when the party in power is willing to test its authority in a free election. Neither of these conditions exist except again in Israel and Lebanon. Even in those two states the electorate votes for parties rather than men in a sort of "controlled democracy," an antiphrasis in political terms.

Meanwhile, in Iran, the oil companies can only think about that proverb with its Islamic overtones that often is heard here: "Giving alms prevents misfortune."

For oil companies the "alms" are likely to be in the form of higher payments for the petroleum that they might take from Iran. Their misfortune might be to be denied all oil from that country if they do not.

This same story is heard in Saudi Arabia.

VII

The Saudis

Abqaiq, Saudi Arabia, is on desert sands so stark that a rabbit would be visible a mile away, though the animal would not have a blade of grass upon which to chew. It is an oil-company town, and prefabricated and concrete-block ranch-type houses line regular streets engineered to precise plans. Flares of burning gas from wells flicker on horizons. Pipes, retorts and huge cylinders of an oil stabilizing plant form a technological jungle on the edge of town, advertising the reason for Abqaiq's existence.

The town is adjacent to the biggest single oil field in the world, Arabian American Oil Company's Ghawar Field. Early in 1973 it was producing at a rate of 4.5 million barrels of oil a day. In a functional office designed for outdoor men rather than desk colonels, Ali Ibrahim Naimi, 36-year-old deputy production manager, answered a phone during an interview.

"Yes," he said to the distant caller. "We're going all out, running at capacity right now."

Hanging up, he turned and said, "We are doing everything we can to squeeze more production from our operations."

He is slight of build, with coal black hair and the air of one who knows how to take care of himself. He started life in a Bedouin tent in the desert among people who could neither read nor write. Now he holds a masters degree in hydrology from Stanford University.

Oil changed his life just as it is modifying the lives of many other Middle Easterners. Nowhere is that truer than in Saudi Arabia, now the area's biggest oil producer. Overnight, this nation of five to eight million people is being transformed by its oil wealth from a dirt-poor country scrounging a harsh living from the desert into a forward-looking nation under construction.

Nobody even knows the exact population, for no adequate census has yet been made. As much as 30 percent of the population may be Bedouins, those wanderers in the desert who live off flocks of sheep, goats, and camels. The ancestors of every true Arab were Bedouin at some time in the past. When, like Naimi, Bedouins settle in towns, they become Saudi, Kuwaiti, Iraqi, or whatever else happens to be the national label of the particular community.

"My mother is of the Naimi tribe, so that is where I get my surname," he said.

Anneke Finch, his attractive blonde English secretary, entered with several cups of coffee. She wore yellow slacks and a sweater, a costume that might have been taboo in any Saudi town or village. Here, in the busy atmosphere of an oil community, it seemed appropriate.

Over coffee, Naimi talked about his background, of a father who had been a pearl diver in the Persian Gulf, of a mother who had left the tribe to marry, then had returned to the tribe with a son when her marriage broke up. So Naimi's early memories are of playing on the carpet of a goat-hair tent in the desert, with flocks of sheep to tend. His mother remarried when he was still a child. His stepfather, an Aramco employee, put Naimi in school. "I liked it from the first," he said.

Few people are as adaptable as the Bedouins. Discover an oil field in their vicinity and soon there will be Bedouins working as laborers, then as oil riggers and drillers, then as mechanics and equipment drivers. "They are natural mechanics," said Sami Labban, a Lebanese agricultural specialist employed by Aramco. He has worked with several projects aimed at inducing the nomadic Bedouins to adopt a settled, agricultural life.

In the old days, the Bedouins liked to raid villages or neighboring encampments, plundering everything in sight. War was considered a manly occupation and loot was the fruit of victory. The firm rule of the House of Saud in this century has bottled those warlike instincts of the desert Bedouins. They still can be accomplished thieves. However, they have a scrupulous code of honor where their own tribe is concerned. Companies that employ Bedouins seek to induce them to transfer some tribal loyalty to the company. Once they become company men they are loyal there too.

Saudi Arabia is a country that, unfortunately, has had too many "Lawrence of Arabia" books written about it, too many movies of desert sheikhs. So the Saudi of today (who may be the Bedouin of a generation ago) may be a much misunderstood person. Outsiders tend to see the Saudis in terms of Lawrence portraits that are more than half a century old. It is as if everyone living west of the Mississippi River in America were evaluated on the basis of Zane Grey novels or television serials such as "The Virginian" or "Gunsmoke."

It may then come as a surprise to those outsiders to find that the camel caravan has nearly disappeared from the desert. When the Bedouin moves from one camp to another, he does it in a Chevrolet or Toyota truck. At Al Khobar, on the Persian Gulf, there is a new marina and boat club. Saudis whose parents were Bedouins in the desert are purchasing ketches and yawls for Sunday outings on the Gulf. High-rise buildings lift into the sky in Jidda amid mud huts. Young Saudis, perhaps of Bedouin parentage, are piloting supersonic jets in this country's air force.

Education provides the fuel for lighting "Aladdin's Lamp." Oil revenues place that lamp in Saudi hands. Naimi was only ten years old when he obtained his first job as an office boy with Aramco, which has various training programs under which it encourages Saudi boys to finish school as they work part time with the company to augment family incomes.

Naimi worked and studied, and won a scholarship to the American University of Beirut in Lebanon. His eagerness while working during vacations interested Aramco. He was given scholarships to Lehigh University and then to Stanford University in the United States. He returned to a good job in the administrative headquarters of Aramco at Dhahran.

But in America he had absorbed some of the drive that fires Americans when they reach for the executive ladder. He asked for a transfer from a comfortable, desk-orientated job to production.

Here his competitive drive soon made him invaluable. Today he is number two in a field that is responsible for a production nearly equivalent to half the crude petroleum output of the entire United States.

"And I love it," said he. Now married, he has two girls of nine and two years old and a boy of six. "The boy wants to be an airline pilot," Naimi added, with obvious pride.

What sort of ambitions does an ex-Bedouin have when he is on the executive ladder of a major oil company? I asked that question, expecting a modest answer. But Naimi replied, "I want to be the first Saudi president of the company." Then, almost belligerently, he added, "Why not? I'm working hard for it."

This is a country on the go, with thousands of young men like Naimi studying to take charge, or working at jobs which already provide broad authority. It is a country with immense oil revenues, which will zoom steadily upward. This money provides the capital for modernizing the country.

Two of the greatest development stories of all time are now underway in Saudi Arabia and in Iran across the Persian Gulf. Here it is a story of peoples awakening after centuries of sleepy stagnation; of new cities rising from deserts; of irrigation projects transforming arid lands into green gardens; of new harbors welcoming giant tankers; and of tribesmen attending school and then graduating with degrees to operate computers, satellite communications systems, jet planes, and chemical processes.

In Jidda, the Red Sea port that is Saudi Arabia's diplomatic center, Anwar Ali, governor of the Saudi Monetary Agency, provided some statistics concerning that revenue. He is a Pakistani of urbane manner who looks like one expects a Middle Eastern diplomat to look—dressed in a western suit, always unruffled, quiet-voiced, and features abeam with friendliness. He has a knack for interpreting figures that has proved invaluable to Saudi Arabia. After doing a yeoman job at the International Monetary Fund, he was drafted in 1958 to help reorganize this country's finances. He did it so well that he has been here ever since at SAMA, this country's equivalent of a central bank.

SAMA figures show the oil revenues for specific years:

1965	$663 million
1968	$1 billion
1970	$1.2 billion
1971	$2.3 billion
1972	$2.9 billion

This only tells part of the story. It has been said that little Kuwait floats on oil. Saudi Arabia has reserves of over 150 billion barrels already proved, more oil than Kuwait, Iran, Iraq, Abu Dhabi, and all the other Gulf states put together. Saudi Arabian oil revenues are estimated at $4.7 billion for 1973 and could hit $25 or $30 billion annually in 1980.

What will Saudi Arabia do with its wealth? Even massive development programs will not absorb all the money that will be pouring into the nation's treasury. Already some monetary authorities fear that currencies such as the American dollar might be upset by the imbalances that are being created by Arab oil money.

It is popular today to do some projecting on the basis of current trends. I did this before visiting Ali and concluded that the cumulative incomes of Arab oil lands between 1973 and 1980 may amount to over $200 billion, nearly twenty times the volume of gold held by the United States and a sum greater than the total volume of liquidity held by International Monetary Fund nations in 1973.

"International cooperation on a broad scale will be necessary to minimize the disruptive effects of a massive accumulation of foreign reserves by Arab nations in the Middle East," admitted Ali. But he insisted that Saudi Arabia is well aware of dangers and anxious to cooperate. He added, "We realize it is to our advantage to handle our surplus funds in a manner that does not disrupt the system. Stability is as important to us as it is to the Western world. You must help us by providing opportunities for us to invest our surplus funds."

Ali saw the situation as a two-part task. On the one hand, it is Saudi Arabia's role to produce and sell oil to the Western world so that the latter will not experience energy shortages. On the other, it will be up to the Western world, mainly the United States, to provide the financial climate that will attract Saudi investments.

In any discussion such as this, Israel is apt to intrude into the conversation. Such was the case here. Ali emphasized that one of the best ways to maintain stability in the Middle East would be "through settlement of its principle problem, the Israeli-Arab question." He added, "A reasonable and honorable settlement would eliminate the principle factor of uncertainty in this area. If this problem were solved, the Western world could look for an era of close cooperation between it and the Middle East."

In Riyadh, there are similar stories about energy and America. Riyadh is a once-sleepy city in central Saudi Arabia that is becoming a metropolis. Four-lane Airport Road leads down a tree-lined boulevard into the heart of the city. Magnificent ministries flank the avenue, air-conditioned palaces of steel and concrete, with curving drives that sweep to marble entrances.

On that boulevard one sees the modern side of this country. Further down, near the souk and the Friday Mosque, Riyadh's mudbrick buildings are tossed together like blocks dumped onto a carpet. Sand-colored houses are so densely packed that the old city forms a warren of geometric patterns with no readily apparent streets through them. There are streets, of course—narrow lanes built for donkey trains and pedestrians, but you see few donkeys in Riyadh today.

This city, like the rest of Saudi Arabia, is fast shifting from donkeys and camels to pickup trucks and limousines. Camels are now used for milk and meat. But many people still follow the Arab way of telling time. The day starts at 6:00 A.M. Thus, what we call 7:00 A.M. is one in the morning under Arabic time. Noon becomes six o'clock, and 6:00 P.M. becomes twelve o'clock, when time starts over again. It is confusing to the visitor, but not to the Arab, who may set his watch to Arabic time. But a visitor must always check the time of an invitation to assure that he is not hours off.

Offices of Sheikh Ahmed Zaki Yamani, the country's Minister for Oil, are tuned to western time. The minister's office is on a long corridor into which light streams through squares of stained glass. The wall of one side of his enormous office is of glass, facing a quiet patio that is like a nook of Granada's Alhambra.

He is a broad-shouldered man of forty-three with a trim goatee and dark eyes that stare directly at a visitor. Usually he prefers Arab robes to western garb, wearing the lengthy *gamboz* with dignity. In the oil industry, he has a reputation for being a hard bargainer, with facts readily available to support his arguments. Like most Saudis he is an avid free enterpriser who thinks that individual initiative is more apt to appear when encouraged competitively than when matters are left to the state.

In the desert one brother helps another, but in the final analysis, it is a man's own ingenuity and strength that determines survival. Thus, early, he develops a self-reliance based upon his ability to face the elements without fear. Socialism's seeds have difficulty sprouting on such ground.

"The United States urgently needs our oil, for it is beginning to experience energy shortages," he said. Then, he explained how he would like long-term supply arrangements so that Saudi Arabia may become a stable supplier to America.

But he, like many Arabs, is growing tired of the manner in which America, as he phrases it, favors Israel in the Middle East at the expense of Arabs. Saudi Arabia's King Faisal is setting the tone for a new, tough policy of oil boycotts against the United States over Israel. In the fall of 1973, after the Yom Kippur War exploded between Israel and Arabs, Saudi Arabia made its position clear. Oil, indeed, was to be a political weapon in the Israeli-Arab confrontation. This desert nation joined seven other Arab states in an oil embargo which was aimed at denying Saudi oil to America unless Israel returned occupied Arab lands. Saudi Arabia's production in October, for instance, was slashed by 26 percent from the scheduled level for that month. Yet, even as the boycott came, Saudi Arabia seemed to be hoping for compromises.

Yamani, a Harvard University educated executive, explained. "The United States is the biggest market for oil. We are the biggest supplier. We have every reason for working together for our mutual advantage."

In 1972, Saudi Arabia produced an average of 5.9 million barrels of oil per day. In 1973, the daily rate hit 8 million in May with all signs indicating that by the end of 1974 Saudi Arabia will have the capacity to challenge the 9.5 to 10 million b/d rate of the United States for the oil-production leadership of the world. In a decade, while U.S. output may stagnate, the Saudi rate may rise to 20 million b/d. Whether any goes to the U.S. remains to be seen.

This is a worry insofar as the U.S. is concerned, for the Israel problem has no easy solution. Yet America needs Saudi oil, if not right now, certainly in the future. Meanwhile, all Saudi petroleum production curtailments hurt Western Europe and Japan rather quickly. Such slowdowns do not hurt Saudi Arabia at all, for it already has more money than it can spend, and the oil kept in the ground is like money in the bank.

Yamani talks with the rationality of a man who understands free-enterprise economics, the benefits of mutual cooperation and the advantages of stability over violence. Others in the area sometimes don't.

There is an Arab proverb that says, "Do not order the tree to be cut down which gives you shade." Nevertheless, there are demagogues in this part of the world who do advocate axing the shade trees of oil in order to spite whatever happens to be in their spite bags at the moment, imperialism, anti-colonialism, America, Israel. Nearly every time the Arab League holds one of its talk sessions, somebody is apt to suggest that oil be utilized as a stronger weapon to force the West to adhere closer to Arab political aims.

One of the weaknesses of the Arab is that he often talks a bigger battle than he wages. The louder he talks, the more he believes his own story until wishful dreams blend into reality. In 1967, President Nasser of Egypt probably had no intention of testing Israel's strength with military arms. But he talked his country into a mass hysteria which had many Egyptians believing that Tel Aviv was only a short march down the road.

In one memorable speech, Nasser yelled his defiance of America and its support for Israel. America could drink the Red Sea, said he, his liquid way of saying that America could go to hell. That speech did more to turn the average American against Nasser than anything else he ever said anywhere. When I said as much to a Palestinian at an oil company guest house on Das Island in the Persian Gulf, he showed injured surprise.

"But surely you must realize that this speech of Nasser's was meant for internal consumption," he said. "Subsequently he extended a hand of friendship to America many times."

"In America we have a habit of believing what we hear," I answered.

"The Arab likes to hear what he believes," interjected one veteran oil man, an Englishman who had spent years in Iraq before transferring to Das, a rocky island which is so cluttered with oil installations that it looks like a floating industrial plant.

The hyperbole of the Arab is one of his distinguishing features whether he is encountered in the Buraimi Oasis on the border between Abu Dhabi and Oman, or in one of the sexpots of Beirut. Language is a vehicle of threats, of bluster, of poetic imagery. Until recently, education was for the very few. The spoken word serves as the medium for transferring ideas from one man to another. Arabic with its nuances and its swinging cadences is more than a language; it is a form of music, a means of expression where the sounds mean as much as the meaning of words.

Education is freezing more and more of the younger Arabs into the bookish molds of Western culture. But there still is enough of yesterday left in Saudi Arabia to retain the vividness of the language. Once in Dhahran, one American woman with a suburban mentality emerged from the supermarket not far from Aramco's administrative headquarters. She wore an enormous hat with a brim wide enough to cover a wash basin.

A Saudi friend walking with me whispered: "She walks under her hat."

Another time on an evening in Jidda, a gentleman who was better at promises than performance, profusely declared that he would correct some misroutings on an airline ticket come the morrow. One Saudi lawyer who knew the gentleman watched his departing figure, then turned to me and said: "The words of the night are coated with butter. As soon as the sun shines they melt away."

In another discussion about manners and lack of them in a Saudi home in that same city, one bearded man with the light of understanding in his eyes, said: "You learn manners by watching those who don't have them."

Christ spoke in parables. But then Christ was a Semite, too. It seems to be part of the Arab makeup to speak in parables. A simple sentence sometimes may make a one-line story which will not only answer a question but which provides cause for thought some while after. Once in Riyadh, when pressing hard for an interview with a very busy government minister, one of his aides grumpily said: "A wise man's day is worth a fool's life."

I didn't take offense, even though I wasn't the wise man in that parable. In the interview which followed, the discussion turned to the unhappy political marriage between Yemen and Egypt which culminated in a long civil war. The Minister derided this attempt of President Nasser's to promote unity of the two countries. Said he: "A marriage of paupers only produces beggars."

Later, when President Nasser offered friendship to Saudi Arabia after a long period of animosity between the two countries, another Saudi in Jidda expressed a distrust of Nasser's motives. Then he said: "If you hear that a mountain has moved, believe it; but if you hear that a man has changed character, believe it not."

A Bedouin saying is: "Three things that prolong life are riding horses, being with young girls, and walking in greenery. Three things that shorten life are fighting with other men, being with old women, and walking in funerals."

Hyperbole is a way of life in Arab lands, but Saudi Arabia seldom issues meaningless threats regarding its oil.

Currently, oil is the catalyst for all economic and social development in Saudi Arabia. In 1972, the country was spending an annual average of $2.3 billion on development projects such as new roads, airports, communications systems, harbors, hospitals, schools, and public utilities. By the end of this decade the spending pace is likely to exceed $6 to $8 billion annually. To put that in the right perspective, one must realize that the latter spending volume is substantially greater than the total volume of America's foreign aid for the whole developing world.

The goal is to transform this desert country into a modern state, which is quite a task. In area, this land is equal in size to the United States east of the Mississippi, a country of wide deserts with occasional mountains. In the deep Southwest, green valleys lie amid rolling hills and mountains. In most of the country nothing much higher than a shrub is found anywhere except in the oases that are scattered about barren landscapes.

Already the Saudi development drive bolsters incomes of thousands of the world's businessmen. An army of salesmen, technicians, and advisers from companies and agencies in the United States, Western Europe, and Japan form new international colonies in Saudi cities. The economic overspill reaches from Lebanon to Pakistan.

A government list of developments underway totals 278 printed pages. Some projects are in the several-hundred-million-dollar range. In Riyadh, surveyors chart a new $200 million airport, which is scheduled to be the most advanced in the world. A new $300 million telecommunications network is bringing telephone service to places that have never seen the telephone except in pictures.

Gone are the days when Arab sheikhs poured money into Cadillacs, real estate in Beirut, and the gambling houses of the Riviera. The modern Saudi minister or executive is sophisticated, well informed, and young, usually under forty-five. Only King Faisal seems old in Riyadh, where men of twenty-eight or thirty may control agencies with budgets in the hundreds of millions of dollars annually. Most seem to hold degrees from Stanford University, the University of California, the University of Texas, Cambridge University, or other illustrious schools.

Abdulaziz al-Sagr, twenty-seven, manager of the computer department at the College of Petroleum and Minerals in Dhahran, is a graduate of the University of Texas who speaks English with a Texas drawl, has difficulty standing still when he orates, and talks about the university's computer as if it might be a favored girlfriend.

"Courses in computer technology are required of all students," said al-Sagr. "We have one of IBM's newest-type computers, the best you can get. Come have a look at it."

I did, then proceeded about the campus. The college is on a high bluff overlooking the sanitized company town that is Aramco's headquarters. Miles of desert stretch to distant horizons. Construction debris cluttered the bluff top where a $23 million building program was transforming barren rock into a modern campus.

In early 1973, there were 935 students in attendance at the school, studying to be engineers and scientists for the country's booming oil industry. Many came from villages of mud huts, from families in which no one else could read or write.

"My father can't even write his own name," said Ali Ghamed, twenty-three, engineering student who comes from a small village in the southwest. He says it without condescension. Family ties are strong in Saudi Arabia, and even a father who cannot read or write still is the head of the family, a person to be respected. The father may choose the son's bride, and when that bride comes home, the father may dictate to her as may any older brothers.

"Me and my brother against my cousin; me and my cousin against the foreigner." That proverb applies forcefully in this country in which ties are first to the family, then to the clan, and then to the tribe. Now a new nationalism is being felt by the educated youngsters pouring out from Saudi schools and also from schools in Europe and America. It is a nationalism born of pride of accomplishments rather than of chauvinism. Schools are rising so fast that a new one is being opened every third day as of 1973. Nearly everyone sees opportunities being created, an illiterate people being introduced to the written word.

Arabs have always honored learning, even when they lacked it. "A book is like a garden carried in the pocket," says one Arab proverb. In a Bedouin encampment there usually is at least one man who can read the holy Koran, who will know the poems of Taabbata Sharran, of Umar ibn Abi Rabiah, or of many other Arab poets. Storytelling and oral traditions are strong. Hundreds of epic verses are carried in memories of rugged, hawk-nosed men who might seem to be far removed from the cadences of lyrical words.

Poetry often reveals much of the Arab character, too:

> With the sword will I wash my shame away,
> Let God's doom bring on me what it may.

Though youth is taking over, old customs and traditions are deeply imbedded in the culture. Revenge for a slight, real or imaginary is part of the Arab code. Yet nowhere in the world is the code of hospitality as strong as in Saudi Arabia. No stranger is turned away from the door or from the tent. The coffee pot is always warming for the guest who may appear suddenly. Courtesy is the rule rather than the exception.

Saudi Arabia is a kingdom in the true sense of the word. There is no inclination toward any constitutional monarchy here. King Faisal Ibn Abd al-Aziz al-Saud is a monarch in the traditional sense, ruling according to the Shariah, or Islamic law code. He is the boss of this country's oil and everything else, though he does delegate details to Oil Minister Ahmed Zaki Yamani and to Abdul Hadi Taher, the capable governor of General Petroleum and Mineral Organization. This is a government agency that is promoting industrial, mining, and oil developments in the country. It is usually referred to as "Petromin."

King Faisal is a ruler who gained power through talent more than birth. In the Arab world, the principle of primogeniture usually does not apply in a royal line. More than one person may be eligible for the throne by right of birth. The king, or a family council, may name someone other than the first-born son as king, with the best-qualified sometimes receiving the nod. Abd al-Aziz ibn Saud, father of King Faisal, did name his eldest son, Ibn Saud, as his heir. But he also named his second son, Faisal, as the crown prince. When Faisal proved to be a better administrator than his elder brother, he was named king in 1964 by a family council.

Modern Saudi Arabian history really begins with the Wahhabi movement in the eighteenth century. This was a puritanical, reformist Moslem movement that favored strict adherence to the Koran. Mohammed ibn Abdul Wahhab, leader of this movement, started denouncing religious laxity in central Arabia about 1740. He found favor in the court of Mohammed Ibn Saud, Sheikh of Dariya, an oasis in the region of Nejd.

The sheikh adopted the teachings of Wahhab and gathered disciples of this new sect as fanatical warriors under his banner. Soon he was the most powerful tribal ruler in central Arabia. His descendents followed the strict Islam of the Wahhabis, while warring against Turks who wanted to subdue the House of Saud.

The fortunes of this family rose and fell, hitting a low point just before Abd al-Aziz ibn Saud (1880–1953) became the sheikh. His capital, Riyadh, was occupied by tribal enemies and Abd al-Aziz spent his early years in exile in Kuwait. In 1902 he won control of Riyadh in a bloody raid. Then he revived interest in Wahhabism and developed a new army of fanatically religious followers, the Ikhwan or Moslem Brotherhood.

Bit by bit Abd al-Aziz conquered all of what is now Saudi Arabia. Son Faisal was a trusted general and then an able governor in the expanding kingdom. On September 22, 1932, Abd al-Aziz's realm was officially designated as the Kingdom of Saudi Arabia.

The official Wahhabi attitude had been one of hostility to foreign and non-Wahhabi influences. Conservatives opposed the introduction of the radio, airplanes, motor cars, movie houses, and financial credit. King Abd al-Aziz skillfully maneuvered his country in the direction of modernism. Yet he and other members of the royal family did believe in the basic elements of the Wahhabi doctrine. Even today, Saudi Arabia is the most orthodox of all the Moslem states. Liquor is forbidden. There are no movie houses except for "underground" theaters, which probably show Mickey Mouse shorts and other innocuous movies. Religious police patrol the cities to make sure that Koranic laws are upheld.

Saudi Arabia's oil industry was already developing when King Abd al-Aziz died in 1953. King Saud, a friendly spendthrift and playboy, thought more of a good time than of ruling the country. When he was replaced by the ascetic Faisal in 1964, it was a popular move. I first met Faisal when he was crown prince, and again after he became king.

King Faisal is an impressive figure when encountered in his Qasr ar-Riyaasa Palace in Riyadh. This is a three-story, flat-roofed building of pink concrete, with balconies, wide windows, and a surrounding garden. Guards in brown *abas*, or cloaks, stood at attention, golden-sheathed scimitars in their belts.

The king was conducting a *majlis*, an informal court where any citizen may petition the king to right a grievance or grant a request. Under Saudi Arabian law, the chief cadi (chief justice) appoints the lesser-ranked cadis in the country. Court cases may proceed through courts of these cadis. But the king is the supreme court and final arbitrator. Citizens are not reluctant to make this final appeal, either, or to start an action at the top at one of the king's *majlis*.

King Faisal occupied a gold plush chair, enveloped in the flowing robes of his country. His headdress was held in place by a gold-embroidered *agal*. Mirrors were hung on green plaster walls, lined by several dozen chairs, each occupied by either an aide of the king, a Bedouin fresh from the desert, or a guard. A coffee man, with large brass pot, poured cardamom-flavored coffee in tiny cups, hardly pausing to halt the stream of brown liquid from the long spout of his pot.

King Faisal was a wiry, angular-featured man with a goatee that was almost gray. His thin body seemed fragile, as if he might be observing Ramadan for twelve months a year. His face had a melancholy cast as he listened to the plea of a robed village elder from the mountain fastnesses near Taif. The king nodded his head, turned to an aide, and said something in a low voice that was scarcely audible ten feet away. The bearded petitioner had been petitioning for an agricultural school in his village as an adjunct to the new elementary school.

A youth of not much more than eighteen, dressed in a western-style suit, wanted to know when the country would be allowed to have a public cinema. Apparently, he was the voice of some intellectual youth group.

"So you want to open a cinema? Good. I will give you a license," said the king. "You may then open a cinema whenever you want."

The youth protested. "Let the government open it."

The king shook his head. "The government should not enter into business if it can be avoided."

"But," said the youth. "If I opened a cinema, the religious fanatics would burn it down."

The king nodded. "When people are ready to accept cinemas, they shall be built. You have answered your own question concerning why we have no cinemas now."

King Faisal is a firm believer in free enterprise. A few years ago, the United States Embassy in Jidda offered the king the standard treaty that the United States draws with most developing countries. Under it the United States promises to guarantee investments of Americans who might inject money into the specific country. King Faisal became annoyed when advisers explained purpose of the treaty.

He said, "But we have no intention of nationalizing anything. We believe in free enterprise. Our mere signature on this document would indicate that there is some doubt about our intention to operate a free economy. So I will not sign this. The world will know from our actions that we favor continuation of the free enterprise system."

I met the king in his private office after the *majlis*. The office was teak paneled, with gold-upholstered chairs and a divan. A tall stack of papers rested on a walnut desk. A single white telephone rested on the glass top. A window opened onto a small garden, where grass struggled to survive in the arid climate.

Arab monarchy is based upon tradition that has both a democracy and an autocracy of its own. A monarch or tribal chief rules as an all-powerful lord, but with help of a family council. In a land where cousin marriages and polygamy are common, large families are normal, and everybody in a clan may be related to everybody else. The Saud family numbers a thousand or so, and more if you count in-laws. Most of those under forty-five have university degrees. Many have managerial experience.

A man does not necessarily have to belong to the Saud family to get ahead. Yamani, the very capable oil minister, for instance, does not. Still, with so many in the Saud family, it is not surprising that you encounter the name in many influential offices.

The king listens to his councillors, for that is the tradition of the desert, of Islam, and of Arab culture. Frequently he takes their advice. If he does not, he probably explains very clearly why.

Should he constantly oppose advice, and if the country faces disaster, dissatisfaction could lead to removal of the king. This happened in the case of King Saud in 1964. In the old days, dissatisfaction frequently was expressed through assassination. Even today, assassination is a congenital risk for any ruler in the Middle East.

King Hussein of Jordan has escaped a half-dozen assassination attempts, from an abortive poisoning to a fighter plane attack upon his plane in the air. King Faisal II of Iraq was murdered in a coup in 1958. Wafsi Tell, the capable prime minister of Jordan, was murdered on the steps of the Sheraton Hotel in Cairo. Sharjah's Sheikh Khalid died in a coup in January 1972.

The Middle East ruler lives by a rough, tough code. He needs courage as well as wisdom, a sense of fatalism as well as a strong belief that Allah is on his side. The average citizen has been bred to this same code, and obedience to authority is part of his makeup. He may gripe. He procrastinates. He may pocket a bribe. But he obeys. One Arab proverb says, "If the King at noon-day says it is night, behold the stars." Another says, "Never sit in the place of the man who can say to you, 'Rise.'"

So the Arab is a master at avoiding clashes with the authority directly over him. He knows what he can do with impunity down to the most minute aspect of protocol and what might bring rewards. He is likely to remember to "look first at the face before you give a box to the ears." Democracy American-style is viewed as akin to decadence, allowing rabble to dictate the shape and content of a government. But this attitude might change if a stronger middle class develops. Don't bank too heavily upon this, for the Arab culture goes deep. When you cut it to the bone, you find that the bone is ultra hard.

In Saudi Arabia, justice is harsh, according to the old code of Hammurabi. The thief may lose a hand, the murderer may be beheaded, the political activist may face a firing squad, and few words are wasted concerning the theoretical rights of the defendant. Punishment certainly may not right a wrong, but the belief is that it may prevent a hundred others.

This I reflected as I was being led into the king's private office by one of his retainers. The king talked eagerly of helping his people, of the tremendous development programs that are underway, and of unfinished tasks. He said, "We have the money, but we are short of trained people for the jobs which must be done."

Later, in the office of Hisham Nazer, Minister of State for Planning, I obtained details of some development programs. He occupied a mahogany-paneled office in Riyadh with a soundproof ceiling and an olive-green carpet. He is decisive and has a reputation for cutting through red tape and getting things done. His fingers played with a briar pipe as he talked about the way education is being promoted.

"Education is the key," said he.

Saudi Arabia has free education from kindergarten through university. Exceptional students may be sent abroad at government expense to work for masters and Ph.D. degrees. Not only is everything free—tuition, books, tickets to cultural affairs, and everything else—but upper-class students are paid stipends that average about eighty dollars a month in the country. Hospitalization and medical care also are free for all citizens. In those cases where local doctors cannot handle a particular ailment, patients are flown to Western Europe, Beirut, or even the United States for treatment.

Admittedly, despite employment of numerous Pakistani, Egyptian, Palestinian, and other doctors, there is a shortage. The present five-year development program calls for increasing the number of doctors and medical technicians from the 1970 figure of 3,400 to 6,900 in 1975.

As with every other program in its five-year plan, the government is running well ahead of blueprints, thanks to the enormous increase in oil revenues.

"Money is certainly not a problem with us," explained Nazer. He showed me statistics of the country's climbing gross domestic product. On a chart it looked like a side view drawing of a mountain-climber's path.

The gross domestic product was scheduled to rise from 1970's $4.1 billion to $6.7 billion in 1975. It approached $5.4 million in 1972 and passed the 1975 goal by the end of 1973.

In 1973 there were 600,000 students in school from kindergarten through university. This compared with 414,000 in 1970, and the number is scheduled to rise to 770,000 in 1975.

At Riyadh University, where struggling young trees provide bits of greenery on a sandy campus, Salem Al Melibary, Dean of Sciences, enthusiastically acted as guide for a quick tour of facilities. Modern concrete and glass buildings were crowded with some of the university's 4,000 students. This compares with an enrollment of 2,900 in 1970 and an anticipated 7,000 in 1975. Ultramodern equipment filled tables and benches of the chemistry and electronics laboratories. Ph.D.s from some of the world's most prestigious schools occupied teaching posts. Most of them are expatriates, drawn here by good salaries and a unique educational situation in which staff requests for money are met almost immediately and with few questions.

On the campus of the College of Petroleum and Minerals in Dhahran, Marwan R. Kamal, Dean of Sciences, emphasized an important point. As much as this country needs university graduates, standards are not being lowered in egalitarian style to roll out diplomas that mean little to holders or anybody else.

"Our students must not only compete within Saudi Arabia, they must compete against the world," said Kamal, a Palestinian-born educator who held American citizenship in addition to a masters degree from the University of Minnesota and a Ph.D. from the University of Pittsburgh. "It is difficult to get into the College of Petroleum and Minerals and even harder to get out with a degree."

The business of producing, or else, is in keeping with the Saudi character. The Saudi is not the hardest worker in the world. He does like to delay tasks until tomorrow, or the day after tomorrow, if possible, but he has an appreciation for education and the fruits of it.

A few years ago, when elementary schools were being introduced for the first time in many communities, the Faisal government paid parents to send children to school. It was feared that otherwise parents would keep them home or close to tribal tents to shepherd sheep or to perform other family chores that youngsters have performed since time began in this part of the world. But parents were eager to send offspring to school, realizing that education is the key to a better life for them. So payments were discontinued.

Instead, the government offered student stipends to persuade children to go further in school, to finish high school. Then the opportunity for a college education often induces the student to continue in classrooms for a few more years.

This government may be termed autocratic by critics. Certainly it never will be called democratic in the Western sense. Yet King Faisal has displayed a shrewd knowledge of the way his people think as he has nudged the country toward the modern age. Television is a good example. Islam has never liked any reproduction of the human form, whether in a statue, painting or photograph. This goes back to Mohammed's early religious campaign against the worship of graven images by the pagans of his day. Many of these images, of course, were reproductions of the human form, fashioned in the belief that pagan gods were very much like men.

When the images were smashed, the Prophet made certain that no new ones appeared in the mosques that were built for religious observances, and the opposition to images became a fixation in the minds of many Moslems.

In modern times, as Saudis met Western civilization head on, the taboo against depiction of the human form became less and less strict. Today cameras are sold in shops of Jidda, Dammam, and other cities. Photographic shops advertise quick service. People laugh over pictures of themselves with the same ego uplift encountered in America.

Then came the television age.

Saudi Arabia delayed introduction of television until the mid-sixties. Wahhabi fanatics opposed it, declared that it was sinful. But in the 1960s King Faisal ordered that it be introduced, knowing that it could be an important medium for education of his people. The king decreed, however, that the first programs on the government network should be religious. Imams read the Koran or said prayers on television, pilgrims were photographed on their way to Mecca, and documentaries of various prophets were presented.

It was difficult for the religious fanatics to criticize these programs. Grudgingly, they allowed people to watch television without interfering with them. Meanwhile, Aramco, the big oil company, opened its own television station at Dhahran, ostensibly to televise programs only to its own employees, many of them American.

The Aramco programs were akin to those shown in the United States. Of course, there was no way that programs could be denied to anybody in eastern Arabia who purchased a television set and tuned to the Aramco station. Soon, a coterie of Saudis were eagerly watching for each episode of "The Virginian," "Lucy," and other programs that were popular in America. And gradually, the government station began to run similar programs.

"Wrestling now is among the most popular thing on our television," said Abdulrahman S. Shobaili, 30-year-old director-general of Saudi Arabian Television, a pleasant-mannered and easygoing man with neatly trimmed moustache and the careful and deliberate speech of an elocutionist who realizes the power of the spoken word. Indeed, he did make his start in broadcasting as a radio announcer after graduating from a language school in Mecca. Then he earned a masters degree from Ohio State University, returning to an executive post in the Saudi government's radio-television network.

Saudi Arabia now has five television stations in key cities with a sixth to be introduced in Asir. The latter will be a color-TV station.

Intrigued by the programming, I inquired if Saudi Arabia now has wrestling as a sport. Shobaili shook his head, saying, "The programs are filmed either at the Olympia Stadium in London or somewhere in America."

Still Saudi viewers develop favorites among the wrestling behemoths. They cheer lustily before the set when one of their favorites gains an advantage. They may moan with pain when he suffers any agonies on the mat.

Now the religious taboos are almost forgotten. Obviously, though, television programmers do not press their luck. If a mini skirt seems to be too short in a particular production, there may be a little film clipping. If there is anything at all that might seem objectionable from a moral standpoint, it will not be shown. Moreover, the Israeli-Arab confrontation intrudes, too. Nothing is apt to be telecasted if it presents Jews or the Israelis in a favorable light.

Saudi Arabia is among the most orthodox of all countries in Islam where women are concerned. Women still are expected to veil faces in public, and until only a few years ago, education was an unnecessary luxury reserved only for princesses and daughters of the very wealthy. Women are not even allowed to hold driver's licenses for automobiles, a restriction that makes life difficult for the wives of oil men who might be living in the country.

Tradition has delegated women to a second-class status. A common phrase is "Women are the snares of Satan." The strict Moslem will contend that the Koran explicity relegates women to a position well below that of men. Certain passages do provide this impression. However, liberal modern Moslems are willing to reinterpret these sections just as the Koranic laws against the collection of interest have been modified to meet the exigencies of modern finance.

Since he came to power, King Faisal has been quietly but firmly promoting education of women. He tried to avoid openly challenging the conservative Mullahs and the fanatics of Islam, yet he refused to be intimidated by them when opposition to the education of women did arise.

When a school for girls was established at Buraydah, a sleepy community north of Riyadh, some citizens objected. A delegation arrived in Riyadh, announcing their intention to camp in the city until King Faisal closed the school. In reply, the king dispatched a battalion of troops to Buraydah, and the school opened under the bayonets of the troops. Enough parents sent daughters to the school to form a student body.

"Now the people have requested that a second girls' school be built in their town," said Prince Mohammed ben Faisal, thirty-five, one of the king's eight sons.

Sitting in the living room of his Jidda mansion, not far from the Red Sea, broad-shouldered Prince Mohammed added, "Once the old way of life is changed for the better, people like it, though at first they may have been afraid of it."

Like others of the Faisal family, the prince is a working member of royalty. He played an important part in the development of a giant seawater desalting plant at Jidda, one that has made life tolerable in that city. King Faisal has insisted that his sons obtain university educations and earn their own way. American schools such as Princeton University are family favorites.

Meanwhile, the drive for female education continues throughout the country. In early 1973, 178 communities in the country had schools for girls. The total was scheduled to rise to 300 by 1975. Female enrollments in the present five-year plan ending in 1975 are scheduled to expand by 95 percent as against 55 percent for males.

Still, there are some anomalies. King Abd al-Aziz University in Jidda is coeducational, in that it caters to both sexes. But boys and girls are rigidly segregated. Not only do they attend different classrooms, but boys attend in the morning while girls go in the late afternoon and evening. Thus there is little chance for students to mix on the campus. Moreover, male instructors are not even allowed to face their female classes. Professors present lectures through closed-circuit television. Girls ask questions and receive answers with help of electronic gadgets.

Attitudes toward women are deepseated among many men. In Jidda, Ibrahim Aini, a local merchant, reported he has three sons during a discussion in his hole-in-the-wall shop. Bolts of cotton textiles cluttered shelves and spilled onto one counter which has a pile of merchandise at one end, a tea table at the other. Bearded Aini, a Biblical character in his robes, proudly showed some snapshots of his teenage boys, listed their educational accomplishments.

"Only one wife?" I asked.

"One is more than enough," he said with a nod. "One woman in the house is trouble, two is a calamity. With the young ones in school, I have three."

He hadn't said anything about daughters. "Then you do have daughters?"

"Oh, yes," he said, with an airy wave of his hand. "But they don't count."

Still the old order changes. In the discussion with Hisham Nazer he had talked about education of females and how this is affecting the country. "We are giving educations to so many girls that soon they will be taking over this country," he said in jest.

Nobody expects female militants to appear on the scene before very long. It is enough now to obtain educations, to slowly give up the veil and the long black shrouds that envelop many of the women of Saudi Arabia. It also is a major gain to see jobs opening for women even though these may be heavily weighted toward nursing, laboratory technician work, and other jobs that traditionally have been female in the Western world.

Education, however, might cause drastic changes much faster than people realize. In Jidda, I visited the office of an official of the telephone company. He outlined how the new development program would soon be bringing the automatic dial telephone to many homes in which people never used telephones before, let alone had one in the house.

As he was talking the telephone rang.

He picked up the receiver, listened for what seemed to be a full minute.

"Baleh," he said.

He nodded his head, said again, "Baleh."

There was a long stretch of conversation at the other end, with the telephone executive futilely trying to interject. The way he was yessing the party, I assumed that the caller must have been an important person, perhaps the local governor.

The executive was shaking his head when he hung up. "Women," he said. "Women!"

He must have noted the question in my eyes. "That was one of our subscribers. She was complaining about the service," he said, unhappily.

"A woman? Complaining?"

"Oh, yes," he said. "Sometimes they stop me on the street and want to know why their telephones don't work. I'm responsible for everything."

He shook his head. "Recently, we gave a story to the press about how, with the new system a subscriber would be able to dial directly to Riyadh, or all the way across the country to Dammam. That woman who just called tried to phone her neighbor across the street this morning and couldn't get through. She wanted to know how we expected to get all the way across the country when we can't even get across the street."

Most revolutions have started with the raising of complaints. The female revolution in Saudi Arabia has a long way to go before anyone notices any gunfire. Yet women are beginning to complain, and complain lustily, too. Arab history tells us that during the glorious days of the Arab conquests, Arab women often rode with their men. In battle they encouraged their men, taunting warriors when they wavered. Modern education may yet draw women from the sociological prisons to which they have been condemned for centuries.

Meanwhile, women are protected by the strong sense of family found in this country. Large families have a sense of cohesion and loyalty which now is rare in the Western world, although once common there, too.

In the old days in Arabia, family ties had to be strong. This was a harsh, dangerous country where each son meant another rifle for a tribe's raiders or defenders. Might meant right in the desert. The family stuck together in order to live.

Today, Saudi Arabia is perhaps as safe as any country in the world insofar as human dangers are concerned. The House of Saud maintains security and stability. Family ties now mean close social relations rather than mere security. But if one man injures or kills another, he is striking at the whole family of the unfortunate, and blood feuds may result. Sometimes these may be settled through payment of "blood money" to the family of the injured party, in a way similar to court suit settlements in the Western world after an accident.

It is common for a father to help a son, and for a son to repay the father when the latter is old by providing the old man with a home and care. If a father is in business, he may educate his son in the intricacies of high finance through the school of experience. At fourteen or fifteen, the boy is given a sum of money. The amount depends upon the circumstances of the family.

The boy is encouraged to handle this money as his own, investing it or using it as capital for whatever venture seems appropriate. The father, of course, is on hand to advise the son, explaining how to take advantage of profit opportunities or warning against potential dangers. But the money remains the boy's, providing him with an incentive for paying close attention to ways of increasing his capital. By the time he reaches manhood, he may have acquired an extensive knowledge of business.

This practice is followed on the Gulf, too, where people have been merchants and traders since long before the days of Sindbad the Sailor. Some say that Gulf dwellers need no training in the acquisitive talents of commercialism. They are born with these talents, just as the camel needs no training to survive the desert.

VIII

Gulf Dwellers

From the balcony of the room in the Carlton Hotel, Dubai's harbor spread before me, a panorama of a busy Arab dhow port where Sindbad the Sailor would have felt right at home. Several dozen of those dhows, or "booms" as they are called here, lay at anchor in the quarter-mile-wide "creek." Smoke fires of cooking braziers rose from poop decks. A Baluchi vessel, its lateen sail furled and a cargo of cattle on deck, chugged upstream on its engine. Auto traffic flowed along the promenade that followed the creek.

Ali Osman, a friend, had insisted on meeting me in his suite rather than in the lobby of the hotel. I had mentioned that I wanted his opinions about the local government and about the future of the Gulf. He was a talkative smuggler who seemed to know everything that was happening up and down the Gulf.

"Won't the discovery of oil here put you out of business?" I asked.

He shook his head. "There is only one business older than mine, and I'm not a woman so I'm satisfied with second best." He laughed. Like many egotistical men he thinks himself more humorous than he is. Then he added: "Oil is a business of risks. So is mine. Both businesses will be a part of the Gulf for a long time yet."

He is a big man with a broad chest which fills his *dishdashah* when he has a sash around his waist. Occasionally, he wears a curved *khanjar* in his belt, probably more for show than because he ever expects to use the sickle-shaped weapon. He has a vain streak which had brought us together. He is proud of being a smuggler, of being in a trade where he matches his cunning against the efficiency, or lack of it, of police and customs officers in India, Pakistan, and Iran. He talks freely about his exploits in the past, but never about ventures underway or planned.

Often in the West people believe that a conscience speaks as loudly in others as it might in themselves. In Arab lands there seldom are guilt pangs concerning circumvented government regulations, as long as one is not caught. Avoiding taxes or customs levies is as much the duty of a citizen to himself as paying them might be to the stout Christian. The Arab believes above all in the Shakespearian adage "to thine own self be true." The very word for taxes in Arabic, *mathalin*, means "inequities."

Still, smuggling didn't make sense to me. Oil revenues pour into the Gulf, providing opportunities for profit in countless ways for shrewd merchants. So I said: "With all the oil money around, it seems to me that this would offer the most profit potential to you."

"Oh, I take any profit opportunity which comes along," he said. Then the tone of his voice changed as he added: "You make the mistake many Americans do. You think that oil and its money will change Arab nature. It won't. Instead, you are finding that it is the Arab who is changing the pattern of oil."

Smuggling is not a crime in the free port of Dubai, one of the world's great entrepôt centers, and the Dubai government disclaims responsibility for the traffic. For centuries, smuggling and piracy have been practiced by seagoing merchants of what used to be called the Trucial States. British warships ended the piracy in the nineteenth century. The truce dictated to these little sheikhdoms under the guns of the warships gave the states their name.

One discovers that little Dubai, with a population of about 75,000 in the city and near-empty spaces behind, is the second largest importer of Swiss watches in the world. In some years it imports a fifth of the Western world's gold production. It competes with the Soviet Union as a silver exporter, though there is no silver mine anywhere within a thousand miles.

"We have a free port," said Sheikh Rashid Bin Said al-Maktoum, ruler of this little sheikhdom since 1938. "Gold and other goods enter here legally. Cargoes depart legally. What happens to these cargoes after they leave here is not our affair."

Sheikh Rashid is an oil man now as well as the chieftain of one of the world's freest ports. Continental Oil Company, in New York, is the manager of an offshore concession that is expanding production to 300,000 barrels per day. Ownership is divided as follows: Dubai Marine Areas, Ltd., 50 percent; Continental, 30 percent; Texaco, 10 percent; Sun Oil, 5 percent; and Wintershall A.G., 5 percent. Dubai Marine is divided equally between Compagnie Française des Pétroles and Hispanoil.

To meet the Sheikh I had solicited help from Mehdi al-Tajir, the chubby government major-domo, ambassador, and bureaucrat extraordinaire who was described to me as "the man who gets things done around here." Al-Tajir is a Bahraini-born Arab who knows how to strike at the heart of any problem in short order. He also likes all of the rewards of success.

Over tea in a marble-floored room in his Gulf-side palatial home, he assured me that arrangements would be made to see the ruler. The place was literally an Arabian Nights palace. A swimming pool with a glass wall faced a sunken bar. One could sit and drink and watch any maidens who might have been swimming in the pool, for a performance that beat television. In the dining hall, fountains trickled in the center of a marble dining table. One room contained dozens of rare Persian carpets stacked about two feet high, warehouse style.

"Collecting them is my hobby," said al-Tajir casually, as if carpets might have been old magazines instead of pieces worth thousands of dollars in some cases.

Al-Tajir is Sheikh Rashid's kind of man and vice versa.

The sheikh, too, is a man who gets things done. He runs Dubai as a business concern, with shrewdness and foresight, handling it as an enterprise which must show a profit. He is chairman of the local telephone company, on the board of the local power company and a penny counter where government money is concerned.

He is small-boned and wiry, with a hook nose, a tuft of graying beard, and a mind of his own. When Arab nations decided they needed a supertanker drydock, the sheikh greeted the idea enthusiastically. Dubai was just the place for it, he said. When the other Arab countries voted to build a jointly financed drydock in Bahrain, Sheikh Rashid refused to go along. Dubai would have a drydock if it had to build the facility alone, he declared. In 1973 it started a $162 million project for the construction of two berths for handling tankers of 500,000 tons deadweight and one for vessels of up to a million dwt. Meanwhile, in Bahrain, a $100 million drydock for vessels of up to 350,000 dwt, has been started.

Dubai itself is a contrasting city of mudbrick houses and modern high-rise concrete buildings, which reflect the prosperity of this city's merchants. Shop windows are jammed with Sony radios, Hotpoint appliances, Remington electric razors, and goods of all types. Automobiles crowd narrow streets. There is a bustle of activity in the local souk. A new $60 million port gives this little sheikhdom the finest general cargo harbor on the Gulf.

One of the Sheikh's British advisers said, "If Sheikh Rashid had been born in New York, he would be heading a big corporation." Then he added, "You know, the Arabs of the Gulf have always been businessmen and traders, people with a sharp eye for the profitable deal. Here in Dubai you find some of the sharpest traders you can find anywhere. A Jew would go broke trying to make a living among them, even if the political element were not a factor."

This was the sort of racial remark that Englishmen sometimes make off-the-cuff. I have heard the same remark made about the Lebanese and also about the Armenians. The Middle East does have its share of sharp traders; the Gulf probably has more than its share. To hold power, a sheikh must be shrewder than most of his subjects and able to control the others.

The sheikhs and kings of this area have always played an important role in the development of Middle East oil—the Shah of Iran, the King of Saudi Arabia, the sheikhs of Kuwait, Qatar, Abu Dhabi, Dubai, and Bahrain. Recently, a dictatorial army officer, Colonel Mu'ammar el-Qadhafi, who rules like a king in Libya, was added to the list. These men have been the powerhouses on the Middle East side when petroleum companies dickered for concessions, for revenue-sharing contracts, and for anything else having to do with oil in this area.

There are advantages for oil companies in this arrangement in that only one man, or his aide in the oil ministry, makes all key government decisions. The companies need not bargain with individuals for oil-drilling rights. Ecologists and environmentalists do not raise voices against sheikhs or kings, if the former even exist. In fact, refinery pollution bothers nobody who counts, for there are no Ralph Naders active in the Arab world, even though Nader is of Lebanese extraction.

Autocracy has its merits when one happens to be on the side of the autocrats. The system enabled the oil companies to operate huge fields, some as large as smaller American states, as single units without worries about state or regional regulations. If any problem arose, the ruling sheikh or his aide eliminated it, especially if the difficulty threatened revenues. There usually was no congress or parliament to debate endlessly over some fine technical point that might cause delays, nor any court to which opponents, if any, might appeal to fight the oil company.

In short, oil companies have been partners of the states, a fine arrangement for the companies when they were the senior partners. When business advantages were coupled with the richest reserves in the world, it is evident why oil production in the Middle East has skyrocketed. The one producing country in the Middle East without a sheikh, a king, or a dictator is Iraq. It is ruled by a military junta that has not always been united, and the oil companies there have had more trouble than in all the rest of the Middle East.

This does not necessarily mean that sheikhs and kings provide the best government for everybody. It might indicate that they do offer the best governments for oil companies. But what happens if a sheikh or a king is overthrown in a coup? The question haunts oil executives in New York and in London. Iraq's example is one the oil companies do not want to see emulated.

Still, the sheikhs and kings have displayed remarkable staying power in a period in which vast changes are underway. Even when a sheikh or a king has lost power, it usually has been to another sheikh or a king, as was the case in Saudi Arabia, Qatar, Sharjah, Abu Dhabi, and Oman. Moreover, no matter who rules, the chances are the government in power is likely to facilitate oil production, unless...

The Middle East always has an "unless" or a "but" somewhere in its vocabulary. Oil is likely to be produced in volume unless some of these low-population countries accumulate such an enormous amount of money in banks that they simply cannot spend it. What do you do when you have all of your citizens living on a cradle-to-the-grave welfare basis? Is it not then better to save the oil in the ground for future sale rather than sell it now for cash, which may do nothing in a bank? The steady rise in value of oil in the ground may exceed interest gains from the money; moreover, the money might decline in value.

Already, two countries, Kuwait and Libya, have reached the point where production limitations appear more attractive than any drive to increase output continually. Eight states have used oil cutbacks as a political weapon. This has enormous meaning to the United States, Japan, and Western Europe, for higher petroleum production is in their interests. Production curtailments are not.

In Kuwait, Abdel Rahman Salem al Atiqi, minister for petroleum and finance, made his position very clear in an interview in his office. Kuwait is an extremely rich little country, but you would never know it from the appearance of the rather run-down Oil Ministry Building in the center of town, just behind the glittering magnificence of the Kuwait-Sheraton Hotel.

Atiqi occupied a large office on the ninth floor. A model of a dhow about four feet in length in full sail on a table was beside a model of a modern tanker about three feet in length. A huge map of the country on one wall showed all of the oil-concession areas and the producing oil fields. Atiqi sat behind a desk in a corner, wrapped in a brown *aba* or cloak, headdress on head. Arabs typically, of course, do not remove that *keffiyeh* when inside.

At forty-six, he has filled a variety of positions for his government —a police executive, director of the health department, a delegate to the United Nations, and minister of oil and finance since 1967. His straight, shrewd gaze indicated that he was not easily susceptible to anybody's proposition. A trace of irritability crept into his voice when one of the four telephones at his elbow rang and he answered, but he was affable when he again swung around.

Atiqi explained that Kuwait's oil production in 1972 averaged three million barrels a day. The country had ample facilities for expanding that output substantially, but had no intention of doing so. "We think that a three million barrels daily rate is about right. This provides us with an adequate income and assures long life for our oil industry. There is no point in raising our rate any higher," he said.

Another telephone interruption provided a break. Then he resumed his discourse. "We are leveling our oil-production rate for two reasons. First, to maintain our oil reserves as long as possible. And second, because we do not see any reason for turning our oil in the ground into money which may fluctuate downward in value. The interest rate we obtain may not compensate us for the decline. Meanwhile, we know that because of energy shortages around the world, the value of our oil will go up." A third reason developed in late 1973 when Kuwait cut output further to protest against Israel. Libya was slashing its production earlier. After reaching a peak of 3.3 million barrels per day (b/d) in 1970, output declined to 2.8 million b/d in 1971 and to 2.2 million b/d in 1972. Though Libya is far removed from the Persian Gulf, it is mentioned here because its oil plays a vital role in the energy picture too, especially in Western Europe. Now it threatens to halt all oil sales to Europe.

To put these figures in the correct perspective, it might be worthwhile to give a few rounded-out statistics. In early 1973, the United States was producing about 9.5 million b/d of crude. The output in key Middle East nations and Libya was this: Saudi Arabia, 6.5 million b/d; Iran, 5 million b/d; Kuwait, 3 million b/d; Libya, 2.2 million b/d; Abu Dhabi, 1.2 million b/d; Iraq, 1.1 million b/d; Qatar, 550,000 b/d; Oman, 300,000 b/d; Dubai working toward 300,000 b/d; and Bahrain, 75,000 b/d.

Meanwhile, oil companies note that sheikhs and kings are becoming more and more sophisticated in their bargaining for more and more revenues. This is a refinement of trading instincts that always have been a part of the Semitic character. Three thousand years before Christ, Arab traders were bargaining over bales of frankincense and myrrh. Their dhows carried these precious cargoes from ports in southern Arabia to ancient Dilmun and to Sumerian and Babylonian ports. After the domestication of the camel, caravans formed transportation links between Middle East cities. These caravans bartered and bargained their way over routes thousands of miles long.

Mohammed's strictures against usury only somewhat affected Arab trading practices. Modern Arabs contend that Mohammed was arguing only against the sharp trading practices of his day rather than against interest itself. Most Arabs in business now feel investment interest to be the fruits of business acumen rather than a sinful return.

This still is not true of the Bedouin. Among them, the worst insult is to be called "a retailer" or "shopkeeper." This is worse than saying that a man sleeps with his mother or that his mother is related to a dog. Yet, curiously, the Bedouin may be a sharp trader himself when buying or selling a camel, disposing of his wool clip, or bargaining over young lambs for market. But these are not the main activities of his life, he insists. He sees himself as a sheep raiser and desert camel breeder, a man of the desert rather than of urban markets.

"In the desert, I am always with Allah," said one Bedouin of the al Awamir tribe, whom I encountered in the Buraimi Oasis of Abu Dhabi. "In town the devil waits in every souk."

There also are numerous expatriates waiting there too, whether the souk is in Kuwait, Bahrain, Dubai, or elsewhere in the Gulf. Oil money has attracted thousands of outsiders for the unskilled jobs at the low end of the scale as well as for the managerial posts at the top.

Afzal Hassan, a well-dressed Pakistani, waited in the lobby of Dubai's Ambassador Hotel to meet the members of an American trade mission. He was a key executive of the Hamad & Mohamed Alfuttaim trading empire, which sold 3,000 Toyota automobiles, 2,000 Honda motorcycles, and other items in the Gulf in 1972.

"We hope to pick up some American lines to sell," he explained. "American goods had been priced out of our market. With the dollar devaluations, they are worth handling again."

Pakistani such as Hassan are encountered behind desks in banks, in executive suites of trading companies, and in shops. Centuries of trade and commerce have leavened Arab populations of the Gulf. Today you find barefoot, turbaned Baluchis; bearded Omanis with skirts that look like Scots kilts; Negroes from ports of East Africa; tall Afghans who often are money changers; wild-haired Yemanis who toil as stevedores. Iranians, Iraqis, and Palestinians fill jobs shunned by the local population. Egyptians teach in school systems, American and British technicians man key oil posts, and British bureaucrats serve as advisers to governments.

Kuwait now has more non-Kuwaitis within its 6,000 square miles than it has citizens—346,000 Kuwaitis versus 387,000 non-Kuwaitis.

There are an estimated 100,000 Palestinians alone, most of them in jobs requiring skills and education, which has become a fetish among Palestinians.

Their number in the Gulf answers the question sometimes asked: Why don't the Arabs take care of the Palestinian refugees?

They do, in some ways. Refugee camps in Jordan, Syria, Lebanon, or Gaza are home for around a million and a half Palestinians, though not at all times. Young males may leave camps for anywhere that jobs are available, from West Germany to Oman. The old, infirm, and unambitious, the very young, many women, and the most embittered anti-Israelis remain behind. Sometimes families join workers for new starts elsewhere.

There are so many Palestinian families in Kuwait that the Palestine Liberation Organization has established its own school network there. In 1972, this PLO educational system had 15,000 students in classes from kindergarten through high school.

Today, Kuwaitis are in the novel situation of being a 47 percent minority in their own country. Non-Kuwaitis hold between two-thirds and three-quarters of all jobs. Kuwait has a cradle-to-the-grave social-welfare system for its own citizens and many do not have to work, or not very hard, anyway. Nearly all consider themselves too good to work with their hands. There is no need for it with the helot class of expatriates that has developed.

The government spreads its wealth through liberal prices for land, which only a Kuwaiti citizen may own. More than $2 billion has been disbursed in this fashion. A Bedouin with a hundred-foot lot may sell it to the government for $50,000 to $100,000. The government might build a house on the site, then resell it to the Bedouin for only the cost price of the home. Enterprising Kuwaitis invest land profits in businesses or in some of the buildings of daring design in burgeoning Kuwait Town. And they grow richer. One government official quipped, "Our government is creating a new middle class of millionaires."

The government stakes citizens who want to enter business, extends easy loans, offers all kinds of services. If a Kuwaiti finds nothing else to do, he works for the government. Any Kuwaiti earning less than $5,600 a year is entitled to special aid.

Kuwait is a social-welfare paradise. There are no income taxes, what with oil wealth pouring in a near-endless flow. Telephone service is free. So is excellent medical care and schooling through university. Poverty has been almost eliminated, though a few Bedouins in the country necessitate that "almost" qualification in this evaluation. Sometimes Bedouins move directly from their goat-hair tents into air-conditioned apartments where they must be taught how to flush toilets. Expatriates are second-class residents when all this largess is being disbursed.

In addition to 3 million barrels of oil daily from Kuwait Oil Company, this little state gets half of the 500,000-barrels-a-day production from the Neutral Zone shared with Saudi Arabia. Producers in this zone are Getty and Aminol onshore and a Japanese group, Arabian Oil Company, offshore.

Kuwait has a freely elected parliament, and the ruling sheikh, Emir Sabah al-Salem al-Sabah, is trying to encourage democracy. He describes himself as a "constitutional ruler," and there is a constitution that guarantees press freedom, forbids offensive wars, and outlines the rights of the people. The emir, however, has "constitutional immunity" and is commander-in-chief of the armed forces. So there is no doubt where the final power lies.

Expatriates cannot vote no matter how long they may have lived in Kuwait. Yet there are so many of them that Kuwaitis wonder uneasily if outsiders eventually could seize control. An underprivileged group anywhere may become a focus for unrest.

Kuwait legislation is tailored to keep citizenship from expatriates. Citizenship is confined to residents in the country prior to 1920, to their male descendants, and to foreign women upon marriage to Kuwaitis. For Arabs, naturalization is possible after ten years' residence, but only fifty persons may be naturalized in any one year.

"They want our labor. They want our skills. They do not want us as citizens," one Palestinian resident of twelve years said bitterly as he sat behind a desk in a government office where he was an adviser. He fled Haifa in 1948 when war broke out, leaving home and career behind. He admits he now is well paid and there is an easy nonchalance in Kuwaiti offices. "They are rather lazy," says he. "But who wouldn't be when there really is no reason why they should work hard? As you Americans say, they have got it made."

One story, probably apocryphal, has a Palestinian taking his automobile driver's license examination before a local board.

"When you meet another driver at a crossroad, who has the right of way?" he is asked by the examiner.

"The Kuwaiti driver," answers the Palestinian.

Concern about possible repercussions from such attitudes plays a part in Kuwaiti political postures. In Arab matters it is more Arab than any other nation. It consistently supports Arab nationalist movements, even though some have Marxist leanings. It maintains a broad assistance program in the Arab world, seeking to spread some of its wealth among less-fortunate neighboring countries. Since closure of the Suez Canal, it has been participating with Saudi Arabia and Libya in payments of about $250 million annually to Egypt.

Cynical critics of Kuwait aver that it is seeking to purchase peace from radical Arab nationalist movements. If so, the policy has succeeded into 1973. Its oil flows without interruption except for those dictated by Kuwait as a conservation policy. Iraq, however, caused some uneasy moments in March 1973 when, in a dispute, it temporarily occupied a Kuwait border post and two islands.

Some internal family politicking exists in Kuwait too. What effect this might have on this country's oil politics is unknown. Arab families such as the ruling Sabah family do not air their linen in public. Still the intrigue has tongues wagging in Kuwait.

The Sabah family is divided into two branches, the Salims and the Jabers. According to recent tradition, the emirship was held by the two branches in turn. In 1965 when Sheikh Abdullah al-Salem al-Sabah died, it was expected that Sheikh Jaber, the present prime minister, would be the successor. A family council decided that he was too pro-Nasser and named the present ruler, another Salem, as the emir. Jaber was given a consolation prize by being named the crown prince and prime minister.

This decision apparently has worked well. The Emir is a strong but benevolent chief of state. The Prime Minister runs the government with authority. Oil revenues roll in and everyone seems happy.

Or are they?

In Kuwait, one hears that Sheikh Jaber bin Ali al-Salem, a nephew of the emir, has been bitterly disappointed at being passed over as crown prince.

"If he could topple the present government, he would," one European diplomat said in Kuwait when evaluating the situation. That, of course, does not mean that he can or will. It does show that in the Middle East, family politics may exist alongside the nationalist forces that receive much more publicity in the world's press. Sometimes those family squabbles have far more meaning to oil men than do the nationalists maneuverings.

Many Kuwaitis take pride in their wealth. It was a Kuwaiti who joked about the two fellow citizens who stopped for sandwiches in a cafe in Kuwait town. On emerging, a nearby Cadillac automobile agency attracted their attention.

"That's what I want," one of them said, pointing to an El Dorado on display at the dealership. Reaching for his pocketbook, he started inside.

"No. No," protested the other fellow. "This is my turn. You paid for the sandwiches."

That same Kuwaiti told me another story, which he declared was true. A friend was so intrigued by automobiles that he could not pass a dealership without entering to purchase a new car. Eventually he had a fleet of fifty-five automobiles. Such stories are responsible for the apocryphal tale of the Kuwaiti who buys a new car every time the old one runs out of gasoline.

Sometimes high-ranking Kuwaitis are self-conscious about such stories. They fear that outsiders might view Kuwaitis as profligate spendthrifts. The stories hardly do justice to the government. This was made evident in an interview with Sheikh Jaber al-Ahmed al-Sabah, the prime minister and crown prince, who might have been emir. He is a plump-cheeked, moustached man of forty-four who shows concern for his fellow men.

I met him in the Sief Palace, a rambling structure of another day set beside the sea. A lone soldier in British-type military dress guarded the door, but not with any aggressiveness. Kuwait is a peaceful country unused to the police-state atmosphere found in some Arab nations today. The Prime Minister's long robes swept the floor as he rose to greet me. An air-conditioning unit formed a backdrop of sound to our conversation. Windows overlooked a fleet of Arab dhows at anchor in the harbor.

He was anxious to talk about social welfare, and said, "Kuwait provides free medical, education, social, and other public services for citizens as well as other residents of the country. We are the only country in the world, I think, which provides such extensive services without imposing any taxes. The per capita expenses of such services are considered among the highest in the world. Moreover, the government spends large sums of money on projects and services aimed at raising the incomes of citizens in general. The amounts devoted to education, public welfare and development represent about sixty percent of our annual budget."

Evidence of this spending is easy to see. Yet, though expatriates benefit somewhat, they are excluded from sharing benefits as citizens. Undoubtedly, this causes resentment.

Oman, at the southern end of the Arabian Peninsula, has had both family intrigue in the ruling family and discontent in part of its population. A backward country, it does have a little oil that might help reduce backwardness provided that oil could be developed further. It has had troubles, though.

For years, leftist Chinese-supported rebels in Dhofar province have tried futilely to carve their own state from Oman. Then, in July 1970, the ruling Sultan of Oman, Said bin Taimur, was overthrown by his son, Qabus, in a palace coup. In the three years since that time, Sultan Qabus has opened his country to further oil exploration and has brought about more changes than rebels were able to effect with their years of violence. Moreover, he has managed to extend his control in the rebel province of Dhofar, thanks to help from Britain and Iran.

Oman's estimated 4 billion barrels of oil reserves are small compared with Saudi Arabia's 150 billion plus barrels. Still, in a world that uses ever-increasing amounts of petroleum products, every bit counts. This will be especially true in the decade ahead when the United States energy gap becomes more pronounced.

The Dhofar rebels, led by the revolutionary Front for the Liberation of the Occupied Arab Gulf (FFLOAG), provide a glimpse of the sort of radical movement that oil men fear. It is violently Maoist—

anticapitalist, anti-oil company, and anti-West. Oman's operating oil companies undoubtedly would face a hard time were this group to take power.

The key company in Oman is Petroleum Development Oman, which is 85 percent owned by Royal Dutch Shell, 10 percent by Compagnie Française des Pétroles, and five percent by Participation and Explorations Corporation, the Gulbenkian Group company. It operates an onshore concession, which, in early 1973 was producing at a rate of 283,000 barrels a day.

Wintershall AG, the West German company, heads a consortium that has an offshore concession. Wintershall owns 59 percent, Royal Dutch 24 percent, Deutsche Schachtbau und Tiefbohr 10 percent, and Partex 7 percent.

Qatar (pronounced "Kutar"), a peninsula in the Gulf attached to the mass of Saudi Arabia, has been involved in a family squabble for control. The present ruler, Sheikh Khalifa bin Hamad al-Thani, seized power in February 1972 from his cousin, Sheikh Ahmed. It was a bloodless coup brought about because of the profligacy of the previous ruler. Already the new sheikh has started to modernize the state and to spread the oil largess a little more equitably among the population.

This is a country that has treated its Al-Thani ruling family very well. The five hundred or so male members of this family have been drawing large stipends from the government since birth as their royal due. Many stories heard about free-spending Arab oil sheikhs probably could be traced back to Qatar.

Sheikh Khalifa has a reputation for being one of the tightest men with a riyal or a pound sterling in the Gulf, where public money is concerned. Like Sheikh Rashid of Dubai, he also is a capable business administrator who understands how to organize a commercial operation to make it viable.

"In this part of the world, you do that by keeping other hands out of the till," one English businessman said in the lobby of the comfortable Oasis Hotel in Doha. Across the harbor a tanker took on a cargo at the end of a long dock. A cool wind rustled the palms of the small garden behind the hotel. A television set in the lounge was showing a Mickey Mouse cartoon, much to the enjoyment of a half-dozen robed guests.

The Englishman had arrived in the capital a week before to conclude a deal for some oil-industry equipment. He had been working his way progressively through bureaucratic layers, hoping to conclude his business in another week. "I'm getting a little tired of watching old cartoons on television," he explained.

Doha, the capital, is a combination of mudbrick and flat-roofed houses, modern palaces, and government buildings of graceful design. Qatar has an area of approximately 4,000 square miles, and a population of 130,000. About 100,000 of these live in Doha, which was a sleepy fishing village only twenty years ago.

The main shopping street is Electricity Street, so named because the town's first electrical power generating plant was located on it. Shops mushroomed along the avenue and now it is a favored shopping area.

Qatar started producing oil commercially in 1949, making it one of the older producing countries in the area. In 1972 it averaged an output of 475,000 barrels a day with production nearly evenly divided between an offshore operation handled by the Shell Group and an onshore Dukhan Field operated by Qatar Petroleum Company. The latter is owned as follows: British Petroleum Co., Royal Dutch, Compagnie Française des Pétroles, and Near East Development Corp., 23¾ percent each, with Gulbenkian interests holding the remaining 5 percent. Near East is owned jointly by Exxon Corp. and Mobiloil Corp. Qatar Oil Company, a Japanese-owned company, has just started offshore production.

Sheikh Khalifa, a short, goateed man with a cheerful, outgoing manner, was oil minister as well as prime minister and minister of finance before seizing power as ruler. He knows how to read an oil-industry contract, and understands reasons for some of the complex phraseology that defines the rights of companies and their obligations to the state.

"Our first priority should be to diversify our economy so that we are not so dependent upon oil," he said.

Ali Jaidah, the shrewd director of Qatar's Department of Petroleum Affairs, emphasized, however, that the bulk of Qatar's revenues would be coming from oil for some while to come. Shuffling papers on his desk to assemble statistics, he said that the country's production rate in early 1973 was close to 550,000 barrels a day, with the rate climbing steadily toward 600,000 b/d.

Because of the world energy gap, potential buyers of that oil are figuratively knocking on the oil department's door. Said Jaidah confidently, "We could sell our oil several times over." Prior to the participation agreement, of course, the producers of the oil here also merchandised it. Now, an ever-increasing share of the oil will be marketed by Qatar, a story being repeated throughout the Middle East. Most of the nations have created national oil companies to handle the new marketing function.

Sharjah, a postage-stamp-sized country on the lower Gulf, also was the scene of a power struggle not long ago, but driving through Sharjah's dusty main street with its dilapidated shops, I found it difficult to see why. The street is lined with mud huts and shops that display cotton textiles, rice, pots, pans, leather goods, and a variety of other products in open-air counters. The silted harbor facing the Persian Gulf exudes a fishy odor that never seems to go away even when the wind is blowing from the shore. Scraggly palms provide the only green in a tan desert landscape. The whitewashed palace of the sheikh with its guard towers and crenelated walls is straight out of *Beau Geste*.

A *Beau Geste*-type scene did occur there in January 1972. Sheikh Saqr, who had been deposed in 1965, led a coup to overpower the ruling Sheikh Khalid. The latter was killed in the engagement and Sheikh Saqr sought recognition as the new ruler.

This was not to be, because Sharjah's 32,000 population refused support. The rebels were seized and incarcerated on an island in the Persian Gulf, where they still are held. Sheikh Khalid was succeeded by his brother, Sheikh Sultan.

Richest of the little sheikhdoms at the lower end of the Persian Gulf is Abu Dhabi, a desert land smaller than Connecticut, which has 90,000 people and an annual income of over a half-billion dollars. Its ruler, Sheikh Zaid bin Sultan al-Nahayan, a big, bold man with a tuft of beard, succeeded to power through a family-supported coup, too.

His older brother, Sheikh Shakhbut bin Dhiyab, had occupied the power seat from 1928 to 1966. A miserly ruler who resented changes being brought to his realm by the discovery of oil in the 1950s, he hoarded oil revenues as they arrived. One report says he bought gold bars with the money, stacking them under his bed until the bed became too hard for sleeping.

"I don't know about that," says one bank official in the Gulf. "I do know that when he was finally persuaded to bank his money, he used to give notice to the bank once in a while that he wanted to come down to count what he had. Calls for cash would go up and down the Gulf, while the bank procrastinated. Then, when the sheikh did show up, there would be a pile of cash waiting for him to count."

The Sheikh's eccentricities finally became too much even for his family. He was deposed in August 1966 after a family council, and Zaid was designated as the ruler. Zaid had spent twenty years as the lord of Buraimi, a lush oasis located on the Oman–Abu Dhabi border.

Shakhbut had given Zaid little money from the government coffers, yet Zaid had spent the money wisely. He improved the irrigation system that distributed water from springs to farms, installed an electrical plant for Al Ain, the main town, and fostered development of an experimental farm.

Zaid had sworn an oath to his mother that he would never plot against his brother, Shakhbut. Eleven of fourteen of Shakhbut's predecessors had been assassinated—reason enough for a mother to worry about quarrels among her brood. Zaid kept his promise, and when offered power, he refused it. Only when his brother was out of the country on his way to exile in Iran did he accept the position.

"With him a promise is a promise," one of his aides said admiringly as we sipped cardamon-flavored coffee in the lobby of the Al Ain Hotel on Abu Dhabi's promenade. He related how Saudi Arabia had claimed Buraimi Oasis for years. "The Saudis offered Zaid a bribe of ten million pounds sterling to swear allegiance to King Saud. Sheikh Zaid was almost penniless at the time, but he rejected the offer and called on every man in the oasis to defend it."

Zaid is an outdoor man, proud of his Bedouin connections. At Bedouin celebrations, when lean men of the desert gather at an encampment, Sheikh Zaid may join in the dancing, twirling a rifle decorated with scroll work of polished silver, his voice joining in the rhythmic chants and shouts of participants. He rides spirited stallions and handles a camel as if he might have spent his life in lonely desert camps.

"He is one of us," a Bedouin told me in the Buraimi Oasis.

In some ways Zaid is a composite of the tug between town and country which is found today throughout the Arab world. The modern world is an urbane society, where success lies not in the remote mountain or desert valley but in the concrete-and-steel canyons of cities. So villagers, country dwellers, and nomads flock to cities such as Cairo, Beirut, Damascus, Baghdad, Kuwait, Dammam, Tehran, and Abu Dhabi town. Often they crowd together in humble homes where the dawn cry of the muezzin may be a call to work as well as to prayer. And they wistfully remember carefree camps on the desert, or of loafing under date palms in a village square, forgetting, of course, the monotony and the missed meals. So they try to retain their old habits and customs even as they hazily learn new practices and conventions attuned to life in urban societies.

They feel that perhaps they are losing something worth saving as they adapt to town life. Yet they know that the old ways cannot be retained in a modern society, at least not in their ancient forms. This causes psychological schisms and religious cleavages, while increasing those suspicions of outside influences that might be forcing some of these changes. The way of the Khalaf, or liberal interpretation of the Koran, may be gaining ground over the Salaf, the literal interpretation that leaves little room for modern philosophical thought. Yet it is not a victory that is coming easily.

Shakhbut's fault was that he could not adjust to this century. Zaid realizes that the adjustment must be made, but he has the romanticist's attachment to the manly qualities and the free life of the desert.

In one declaration of policy, he said, "Our material progress will do more harm than good unless accompanied by social progress serving as the foundation for a civilization in which we can create a society based on the true Arab tradition, a society which can enshrine and preserve these traditions for the other Arab countries."

That statement reads as if it might have been written by one of his British advisers, even though the meaning probably came straight from the heart. In one interview he put it to me much more feelingly in the parable form that fits his character better than the neatly phrased semantics of the speech writer. "Our money is like the abundant water which comes down from the mountains in rivulets. This water can run to waste in the desert sand and do no good. But if the water is channeled into a canal which leads to farm fields, it can create a fertile land. We must channel our expenditures so that the spending flow brings good to the people."

I met the sheikh in his old palace in Abu Dhabi town, on an island connected to the mainland by a causeway. It was only a fishing village before the discovery of oil. Now it is one of the busiest construction sites on the face of the earth. Soaring oil revenues stimulate construction of new port facilities, glittering new government buildings, high-rise apartments, new hotels, and wide avenues. Everywhere there is construction debris. From morning until night, pile drivers are pounding, welding torches are hissing, and Yemeni building workers are singing their chants.

The old palace is another of those *Beau Geste* forts with whitewashed walls and crenelated walls and towers. I met the sheikh in the audience chamber, a high-ceilinged room about sixty feet in length, with chairs and divans along sides. Sheikh Zaid stood up, a tall man with the build of a heavyweight boxer who has kept himself in trim. He wore black robes with gold edge, a white headdress on his head. After shaking hands, he motioned to a seat beside the stuffed chair that was serving as his "throne."

When I mentioned that he probably was experiencing difficulty in finding ways to spend all the money pouring into his country, he said, "Better to have the problem of spending it than the problem of finding it."

Wryly, he said: "We have much money—people tell us."

People are flocking to Abu Dhabi to profit from it, too, said the sheikh. "We attract people who want to sell us things, like the honey pot attracts flies."

His humor may be very practical at times. Earlier, an aide had told me of one memorable meeting of Abu Dhabi's Council presided over by Sheikh Zaid. The meeting began solemnly enough, with everyone arriving in his Arab robes to be greeted in chest-hugging fashion by a sheikh who seemed to be in better humor than usual.

Once the meeting started, first one, then another of the council members was afflicted with a maddening itch. Soon all but the Sheikh were furiously scratching.

About that time, the Sheikh burst into loud laughter. During the round of greetings and hugs, the Sheikh had surreptitiously sprinkled itching powder on robes of his assistants.

Like most Arab rulers, he depends upon his council, though he might subject members to humorous indignities at times. When faced with a problem, he is apt to discuss it with aides. Then he may slip behind the wheel of his high-powered Mercedes, drive into the desert, and park beside a dune. He relaxes in the seat, ponders the problem awhile, then returns to his office with an answer.

One of his development projects has been a four-lane superhighway between Abu Dhabi and Buraimi Oasis, ninety miles into the desert. This is certainly one of the most unique highways in the world, a beautifully engineered road that is usually barren of traffic except perhaps for a goat or camel which may wander onto the pavement from some Bedouin's pasture.

The sheikh also has a 45-foot launch. He likes to take it into the Gulf, don trunks, then go over the side. One aide said, "He swims so well that he could have been a pearl diver."

Pearl diving still is practiced in the Gulf, of course, though divers are finding that they can make more money working for an oil company. So the pearl fleets that once provided most of the income for ports like Abu Dhabi, Kuwait, Bahrain, and Dammam grow smaller every year.

One British adviser said of Sheikh Zaid: "Of all the Arab leaders he probably comes closest to the American concept of the desert sheikh." The sheikh has been especially partial to the British ever since they supported him against the Saudi Arabian claim to Buraimi Oasis. He seems to favor Americans, too, for their drive and ability to get things done.

"We need more of that here," he admitted. Then he shook his head in wonder as he recounted his experiences on his only trip to New York City when accompanying an ailing brother to a hospital for special treatment. "Everybody in New York runs to work." He shook his head in amazement, then repeated, "Runs!"

To get the oil out of the ground Abu Dhabi has had others running for them. Like Kuwait, it now has an expatriate population that outnumbers the local citizenry. The latest census showed that foreigners accounted for 56 percent of the total population, with Iranians and Indians the largest groups among the expatriates. This raises some of the same questions which may be asked about Kuwait. But Abu Dhabi, with far fewer skilled and educated people than Kuwait, must import foreign labor to produce its oil.

Abu Dhabi is the scene of an oil boom that is even more extraordinary than that of Kuwait, the country that has been synonymous with oil wealth in the minds of many people. Abu Dhabi has copious quantities of oil both on land and in the seas off its shores. Oil companies first displayed interest in the country as long ago as 1935, but World War II delayed development work.

It was not until the 1950s that exploration began in earnest. By 1960 petroleum had been discovered in commercial quantities. The two key companies are Abu Dhabi Petroleum Company, which has the onshore concession, and Abu Dhabi Marine Areas. The former is a combine of Iraq Petroleum Company, Ltd., companies—i.e., British Petroleum Company, Shell, Compagnie Française des Pétroles, Exxon, Mobiloil, and Gulbenkian interests. ADMA was two-thirds owned by BP and a third by CFP. But early in 1973 BP sold 45 percent of its interest to a Japanese group for $789 million.

There is no doubt that Abu Dhabi has oil and more oil, and it is pushing production hard, though this is tempered by willingness to use oil as a political anti-Israel weapon. From nothing at the start of the decade of the 1960s, production has been pushed to slightly more than a million barrels a day in 1972 and to an estimated average of 1.2 million b/d in 1973. By 1975 production is expected to reach 2 million b/d. This would double the country's income to $1 billion from the present level of about a half-billion dollars.

Output in 1973 was augmented by Abu Dhabi Oil Company. This is a group composed of the Japanese firms of Maruzen, Daikyo, and Nippon Mining. Production started at 30,000 barrels a day, with expectation of reaching 100,000 barrels a day by 1976. Abu Dhabi

National Oil Company has been formed by the government to market the oil it will be receiving under the recently negotiated participation agreement with oil companies.

Abdullah Absi, Director of Petroleum Affairs for Abu Dhabi, said in one interview, "We have the king of all the oils in the Gulf, low-sulfur, high-quality crude." He picked up some letters on his desk and passed them over to me as he added, "You can see we are getting letters from all over the world from people who want to buy our oil through our national oil company."

The letters included a few from American groups on the East Coast of the United States. They expressed interest in long-term agreements for Abu Dhabi oil, which could be imported into the United States in crude form and refined in facilities yet to be built. Abu Dhabi would like to join such groups in joint ventures, Absi explained.

There must be a moral of some kind here. Only a dozen years ago Abu Dhabi was a prime candidate for foreign aid but was too small a spot on the map to be even noticed by the United States. Now Abu Dhabi eyes potential investments even in the U.S. if political differences over Israel can be eliminated. Here it is not alone, Saudi Arabia also is interested in investments on like terms. Other Arab nations may join as exporters of capital, sending their bundles of bank notes abroad. Iran is the first Middle Eastern nation to move into the U.S. In July 1973, it closed a deal with Ashland Oil Company whereby it acquired a 50 percent interest in Ashland's refining and marketing operations in New York State.

Despite the expenditure of hundreds of millions of dollars upon development projects, Abu Dhabi will not be able to spend all the money pouring into its treasury from oil revenues. And Sheikh Zaid shows every intention of making sure that this money is not thrown away merely because it is there.

When I expressed an interest in visiting Buraimi Oasis, he said, "You will like Buraimi. There you will find many of our development programs."

He clapped his hands and an aide came running from an adjacent room. A few quick words and we had permission to visit the oasis that was the scene of an international incident between Saudi Arabia and Abu Dhabi in the 1950s. That incident illustrates some of the stresses and strains that exist in the Arab world, with luck, the probable source of much of America's oil in the future.

IX

Gulf Politics

Khaled Fayyad, the Abu Dhabi director of agriculture at Buraimi Oasis, is a small, bespectacled Jordanian with dark moustache, iron-gray hair, and an intense desire to avoid any political talk except that involving his pet hate, Israel.

"I don't know about that," he said, when I asked him about the Buraimi Oasis feud between Abu Dhabi and Saudi Arabia. Earlier, the conversation, mostly one-sided on my part, had involved the squabbles that plague the Gulf, such as Iraq's dispute with Kuwait, the unsettled claims of nations to portions of the Persian Gulf, and Bahrain-Iran differences. Then I had asked about Saudi influences in Buraimi.

We were sitting on the porch of the guest house of Sheikh Zaid at Al Ain, the main village in the Buraimi Oasis. An obsequious servant provided us with numerous cups of mint tea. A late-afternoon sun created long shadows among the forest of date palms. Veiled women in black trooped along the road in an endless stream, water jars atop heads. A donkey train wended its way down a dusty lane among palms, harness bells jangling, a gray-bearded driver nudging animals along with a stout stick.

Buraimi Oasis is a six-mile-wide cluster of villages and gardens, with a population of about 15,000, counting children, who seem to be everywhere. Villages are half-hidden among the date palms, the jagged mountains of Oman a purple barrier behind. Part of the oasis and three of the nine villages lie in Oman, and an open border cuts through Buraimi, unmarked and unnoticed. Omani soldiers with bandoliers of bullets across chests had waved us on without even checking papers earlier in the day when one trip took us several miles into Oman and back again.

It usually is thus when governments remain in the background at a border. When each side feels that it must jealously guard territory, tight pass systems are invoked, incidents occur, and enmities of governments infect populations. Oman and Abu Dhabi are sensible. So the oasis exists as a unit even though it straddles a border.

"I try to keep out of local politics," said Fayyad. "After all, I am a guest in the country." There was an awkward pause. Fayyad preferred to talk about his past. A Jordanian, he had headed Jordan's district agricultural department at Nablus on the West Bank until June 1967. After Israeli overran the West Bank, Fayyad's concern for his family grew. He said: "They let us know that they think they are the masters now." Finally, he bundled his family into the back end of a truck and joined the refugee parade across the Jordan River into King Hussein's truncated realm.

Then he began to talk about the year he had spent at California Polytechnical College when studying agriculture. One sensed that this may have been a happy time, a wondrous period when life offered interesting challenges and when the fire of confidence burned strongly. But I wanted to talk about Buraimi Oasis.

Yosef Mohamed Ali, chief justice of Buraimi province, had no mental blocks. Perhaps a judge must listen to so many intimate tales that his tact buds are blunted. Ali was a smiling, friendly Sudanese, as black as the goat-hair tents of the Bedouins. He had been educated at London University. When he spoke, I needed only to close my eyes to feel myself in Britain, the land of the broad *A* and the precise vowel. "Saudi Arabia contends that its borders include this oasis," he said. "It was the British and Sheikh Zaid who changed their minds. A jolly show that was, too."

The jolly show involved counter claims, Saudi occupation in 1955, and ejection of Saudi troops by a British-led detachment of the Trucial Oman Scouts. This was a small military force that the British had created to maintain order in the area.

Always when one crosses the paths of British colonialism one encounters remnants of native military forces created in the golden days of colonialism. Britain seldom policed its empire when the task could be allocated to locals, whether black, yellow, or brown, though always there were British officers with riding crops and polished boots, exuding that reassuring air that everything would be all right, so why fret.

"It is the oil and the hope for oil which encourages territorial claims," said Ali.

I nodded, though the Buraimi Oasis dispute is more akin to that of feuds before the oil age when the water in a well was cause enough for a battle. But Saudi Arabia's King Faisal is not one to launch thoughtless wars. The Buraimi Oasis dispute has been allowed to recede quietly into the background.

Still, territorial claims are part of the politics of the Gulf, ready to explode with some provocation. It may seem incongruous that one nation should covet miles of sand where borders are unmarked and where a camp at dawn seems no different from a camp sixty miles away at dusk. But lonely desert acquires worth when there might be oil beneath the sands.

Kuwait and Saudi Arabia argued endlessly about their borders. Britain acted as mediator and established the Neutral Zone between them. Meanwhile, two oil companies have discovered oil on the shore, another in the waters offshore. The onshore companies are Getty Oil Company, which was put together by billionaire Paul Getty, and American Independent Oil Company (Aminoil), a subsidiary of R. J. Reynolds Industries. The offshore concession belongs to Arabian Oil Company, a Japanese concern that is owned 80 percent by Japanese Petroleum Trading Company.

This heightens interest in any bit of land or area of the Persian Gulf which has exploitation possibilities. Moreover, in recent years several other elements of instability have been added to the geopolitical scene. The Soviet Union is aggressively attempting to influence the complex political gyrations of nations in the Middle East. Britain, meanwhile, has retired as a military power in the Gulf, though its economic and diplomatic weight still is considerable. Fear of a political vacuum have prompted Iran to seek to fill it before the Soviet Union or anyone else can do so. The Shah of Iran has created the most powerful army, navy, and air force in the Gulf. The Iranian Navy was the first in the world to develop a Hovercraft squadron.

Its air force has U.S.-built Phantom F4 jets in its fleet. The army includes British-built Chieftain tanks in its battle force, a tank that some military men consider to be the best in the world. In 1973, the shah announced purchase over a period of years of $2 billion worth of arms from the United States.

The quarrels of this area may seem remote from America. But the energy gap brings them within the shadow of the Statue of Liberty, or perhaps one should say to the gates of General Motors Corporation. The Persian Gulf contains the richest concentration of petroleum to be found anywhere, about 60 percent of the entire world's known reserves. This would be a valuable prize even if the United States did not need a barrel of that oil. Unfortunately, the United States needs a great deal of it.

In testimony before the U.S. House of Representatives subcommittee on the Near East in July 1971, the U.S. Office of Fuels and Energy of the State Department told Congress that America would be consuming around twenty-four million barrels of oil a day by 1980. At least twelve million b/d would be imported, three-quarters of it from the Gulf and North Africa.

America certainly does have a strong interest in the Persian Gulf, its oil, and its stability. The ambitions of tribal sheikhs and other leaders may have a bearing upon the oil that rides in supertankers to the United States from this part of the world.

The Persian Gulf itself seems designed for political blood feuds. Today oil has become so important that nations jealously claim their continental shelves, extending rights further and further into the oceans and seas abutting shores. The Persian Gulf is a shallow sea, seldom deeper than three hundred feet, and international custom views the continental shelf as any bottom up to twice that depth. Thus theoretically, Iran might claim the whole Gulf as part of its continental shelf. Conversely all the small sheikhdoms and Saudi Arabia might claim the whole Gulf as part of their shelves. So the easy way might seem to be to divide the Gulf right down the middle.

That is not so easy, however. Where should the Gulf start, from the shoreline or from the outer limits of islands that nations own? Obviously, each nation likes to claim that its shoreline really extends to the outermost limit of its outmost island, and would say that the Gulf should be divided from that point.

The problem is accentuated because of the way shorelines curve and indent. Thus, one small sheikhdom's sea claims often overlap another's.

If there were no oil, disputes probably could be settled without too much trouble. It now is evident that the bed of this Gulf covers some of the world's richest oil fields. Countries along the shore have granted concession after concession, and these already extend far from shores. It was inevitable that arguments should develop.

Already two American companies have been caught in the backwash of Gulf bickering, Buttes Gas and Oil and Occidental Petroleum. The former is the operating company in a consortium that has an offshore concession from Sharjah, that 1,000-square-mile spot of geography on the lower end of the Persian Gulf. The consortium and interests are as follows: Buttes, 37.5 percent; Ashland Oil, 25 percent; Skelly Oil, 25 percent; and Kerr-McGee, 12.5 percent.

The consortium struck oil about nine miles to the east of Musa Island, which Sharjah claimed. The initial discovery had a run of 14,000 barrels per day, and by July 1973, there were predictions that a 100,000 barrels per day rate could be reached at the developing field. One source close to Sharjah's sheikh even averred that a 300,-000 b/d rate might ultimately be attained.

Meanwhile, Iran claimed Musa Island. Prior to discovery of oil it had backed its claim by landing troops and had taken control of the islands. The islands dominated the vital Straits of Hormuz which connects the Gulf with the Indian Ocean. Should one vessel be sunk in this passage it might cork the Gulf, preventing shipment of Gulf oil to market. Iran wasn't taking any chances, though it did promise to divide any oil revenues from Musa and vicinity with the Sheikh of Sharjah.

Earlier, Occidental Petroleum Company had obtained an offshore concession from Umm al-Qaiwain, a tiny Persian Gulf state with a 15-mile shoreline, an area of six hundred square miles and a population of five thousand. Its principal export and means of subsistence is dried fish. The Occidental concession overlaps that of Buttes Gas, but Buttes has possession, equivalent to 99 percent of the law in the Gulf. Occidental has futilely taken the case to American courts.

Claims and counterclaims are especially bitter at the head of the Gulf, where Kuwait, Iraq, and Iran uneasily share the coast. For years, Iraq has had designs upon Kuwait, a one-sided love affair that does not preclude political rape if the opportunity arises. In fact, Iraq did make an abortive attempt to annex Kuwait in June-July 1961, when the unpredictable General Abd al-Karim Kassem ruled Iraq.

As in several other squabbles in this area, I found myself on both sides of this one in a matter of days. It was in late June that I arrived in Baghdad on an Iranair flight from Tehran. Ibrahim Wahab, then Dean of the Baghdad University's law school, greeted us at the airport. He is a short, dark-complexioned man with an unruly shock of hair, who showed willingness to go far out of his way to assist a stranger.

He illustrates the innate hospitality in most Arabs, that willingness to extend oneself at the behest of a friend, or even the friend of a friend. I had been given his name, and he mine in a roundabout fashion. A friend of mine knew Chief Justice William Douglas. Justice Douglas had been in Iraq a little earlier and had met Wahab. Through my friend in Washington I had obtained Wahab's name, and Justice Douglas' secretary had written him. In such fashion lasting friendships are formed in the Middle East.

Wahab drove us into town. The road passed low mudbrick houses scattered over rich plains beside the Tigris. Traffic intensified as Baghdad closed about us. People swarmed along streets, pushing handcarts loaded with vegetables, toting bags on shoulders, darting between automobiles on jammed streets.

Nasser Al-hani, Iraq's Director-General of Publications, waited for me in the Foreign Office, a graceful, oriental-styled building with false minarets and decorative arches. There was the usual transference between aides as I worked my way upward through the hierarchy of the Foreign Ministry to the senior official on duty that day. One seldom walks directly into an Arab dignitary's office from outside, even when an appointment has been arranged beforehand. Each official in the bureaucratic layer must show his authority by assuming charge of the visitor, if only briefly. Each has his own style of human censorship, as if one might be a page of a document that must be stamped at each office along the route. The visitor reaches the destination bruised in spirit but he has an intimate knowledge of the physical layout of the building.

Eventually, I found myself talking to Ibrahim Chorbachi, an

undersecretary in the ministry, a man with a high forehead and the ascetic features of an El Greco monk. "Kuwait is an integral part of our country," he said during our conversation. "It is in Kuwait's interests to return to the mother country."

"With all its oil revenues?" I asked innocently.

Chorbachi glanced up, relaxed when he caught my eyes. In any interview one may spar, or bargain, or press hard with the Arab, but must never show anger or partisanship with the opposite side. I have found that the pointed question presented with the bland manner provides much more information than *any* question asked with pointed manner.

"Its oil revenue is only a side issue," said Chorbachi.

"But of course the oil revenue of a daughter should belong to the mother country," I said, bland as before.

"Of course," he said.

At that time, Kuwait's oil revenues amounted to a respectable $300 million a year, a trifle more than that of Iraq's itself. Britain had ended its 62-year-old protectorship over Kuwait on June 19, 1961. General Kassem immediately made his claim for Kuwait. Even as I met with government officials in Baghdad, part of Iraq's 90,000-man army was being massed near Basra, in the Southeast, apparently ready for a quick swoop on what seemed to be a defenseless Kuwait.

Britain quietly let it be known that it would protect Kuwait, if necessary. But nobody in Baghdad seemed to believe that Britain would show any resolve, were it presented with a showdown. "Britain knows better than to take any military action against a progressive Arab regime," I was told by one Iraqi official in Baghdad. "This would be interpreted as a step backward to colonialism throughout the Arab world. Britain would lose far more than it would gain."

With that, I realized I had better get to Kuwait. I took the first plane to Beirut, then transferred to the next flight to Kuwait. It was a nearly empty plane. This is always a sign of trouble in the Middle East, as is an empty street in daylight. I arrived in Kuwait with the first units of the British Army. In forty-eight hours the British were deployed all along the border with their Kuwaiti allies, waiting for the expected thrust of General Kassem's forces.

With the army came about forty or so reporters and photographers from the world's media, all eager to justify their prodigious expense accounts. The first morning, the government called a press conference, sending a fleet of taxis to the Golden Beach Hotel to transport the visitors. Taxis were air-conditioned Chevrolet Impalas and Ford Galaxys, each driven by a white-robed Kuwaiti with a flowing headdress. The temperature was eighty degrees in the early morning.

Major Abdullah Sayed Rejab Rifai, a public-relations officer with the Kuwaiti Army, herded the press contingent into taxis. Horns honking, the cavalcade rolled to G One, rear headquarters of the army, a nineteenth-century fort on the edge of this sprawling city of modern buildings, sandy lots, and flat desert environs. At the gate, two antique cannon angled barrels at a sky already growing misty with dust and heat haze.

In a mahogany-paneled lounge, sandaled orderlies in white uniforms served a round of drinks—orange juice and Pepsi-Cola. An officer detailed positions of forward troops about eight miles from the Kuwaiti-Iraqi border. Then, as a windup, he said, "Today we will go to the front."

Feet stirred. Newsmen smiled and photographers wiped lenses. Everybody ordered a second round of Pepsi-Colas. The "front" was a mysterious line amid sand dunes and hard-packed flats to the north.

The party filed into the courtyard of the army's fort, climbed into taxis. The temperature was ninety degrees in the shade.

Drivers zoomed motors. Taxis rolled onto the main road north, which was a wide strip of asphalt as straight as an outstretched tape measure. Tufts of camel grass grew on the desert, gray splotches that only a camel could identify as grass. The sun glinted with a metallic sheen on an arm of the Persian Gulf to the right, and the smooth, unruffled water looked even drier than the sand.

At seventy miles an hour drivers raced down the road, following a Command Car De Soto sedan occupied by a trio of Kuwaiti Army officers. Each of the taxis sought the number-two position. Cars passed and repassed wildly in the traffic gaps left by supply trucks returning from the front. Nobody wanted to be last, for if the pavement gave out, the last would eat dust for miles.

At a guard post a sentry with an automatic pistol halted the convoy. Nobody had told him that any news correspondents could pass into the potential battle zone. Nobody in an army ever seems to tell a private anything by way of explanation.

Major Abdullah emerged from the lead car with that authoritative manner that military officers strive mightily to exude. A stream of rapid-fire Arabic wilted the sentry. Again the party moved forward. The temperature was ninety-five degrees in the shade.

At a concrete-block military compound the party paused, one of those catch-your-breath stops that seem frequent in any tour organized by the military, regardless of what army. If wars were won by logistics, this would be a world of defeated nations still unraveling the snafus of World War II. This stop seemed to be a pause to count the taxis, as officers worried about strays.

A group of Bedouins squatted in the shade, belts of bullets across chests. Each man carried a rifle, butt resting on the ground, barrel pointing at the sky. They seemed to be part of the militia, perhaps the last line of defense—the very last.

Photographers unslung cameras, and Bedouins posed eagerly, especially when folding money appeared, standing stiffly at attention like wooden soldiers in a toy display. One lensman gestured to an unshaven warrior who knew a little English, encouraged him to "look wild." Half-heartedly, the Bedouin raised his rifle as if the hot sun might be burning its stock. "Hi!" he pipped. The photographer groaned, then snapped a picture.

"Show us what you will do to Kassem," another cameraman asked. The Bedouin obliged with a weak cry and a gesture to match. One dissatisfied photographer swore.

A taxi driver mumbled something in Arabic. The Bedouin drew back, glowered. This time his yell was bloodcurdling, his raised rifle menacing.

Later the driver explained. "I told him that the government really had gathered them here to put them to work."

Inside, in a conference room, Brigadier Mubarek Abdullah al-Jaber al-Sabah, 29-year-old Deputy Commander-in-Chief of the Kuwaiti Army, conducted a press conference. In a red beret, the Sandhurst-trained officer presented a jaunty figure, very British, very proper. Though deputy commander, the brigadier was the highest-ranking officer in the Kuwaiti Army. "The Emir is the commander," someone explained.

Brigadier Mubarek briefly outlined a situation that seemed to concern perimeters, advance posts, defensible positions, and strategic reserves—great stuff for the military reporter, especially one with imagination, but nothing much for photographers and television crews.

He closed with a ringing statement: "We are prepared to defend every inch of our country." The way he said it had pencils moving and cameras clicking. Perhaps there would be a war after all, with no trees to block camera shots.

Brigadier Mubarek Abdullah al-Jaber al-Sabah smiled at one camera, turned to smile at another. Outside, the television news cameramen had him make his statement over again. They hadn't lugged any lights along for interiors. Who expects to find an "interior" on a desert front?

"We are prepared to defend every inch of our country," repeated Brigadier Mubarek, putting even more emphasis into it this time.

"Wait a minute," called Nigel Ryan, lean, wavy-haired interviewer with a British television crew. "Let's have that over."

Brigadier Mubarek's smile tautened. "We are prepared to defend every inch of our country," he re-repeated.

The camera whirred a few seconds, stopped. "Once more please," pleaded Ryan. Brigadier Mubarek's eyes frosted. He took a deep breath, and seemed to be saying a prayer to Allah. Then, he re-re-repeated, "We are prepared to defend, etc."

Another movie cameraman, who had been photographing Bedouins, came running. "I missed that," he cried. "Once more, please."

Brigadier Mubarek scowled, motioned to an aide, who bustled correspondents into waiting taxis. Again the race resumed behind the command car, only now there was no road, just flat, hard desert as smooth as a two-mile-wide airstrip, with boulders haphazardly on its surface. Drivers stepped on accelerators. Everybody drove in second place behind the command car, riding like a motorized cavalry charge in a twenty-first-century version of a twentieth-century Western movie.

Ahead loomed a low sand ridge. Gun emplacements appeared on the hillside. Tan pup tents hid foxholes. A tank gun poked above a dune, hidden under a tawny net. British soldiers, bare-chested and in shorts, waved to the onrushing taxi fleet. The temperature was 105 degrees, but it was comfortable in the air-conditioned taxis.

Somebody in one car sighted a half-dozen English soldiers pushing an antitank gun onto a terrace dug in the hillside. Drivers jammed on brakes, and cameramen tumbled from cars to photograph the first bit of "action" encountered.

The gun crew pushed and shoved. Inch by inch the gun moved up the sandy hillside as yards of film rolled through movie cameras. Correspondents and photographers gathered around and watched. The corporal in charge swore, one of those great army oaths that seems to echo even in an open lot. Taxis honked. The newsmen rushed to vehicles, and off we went again. The temperature was 110 degrees.

Now there was plenty to photograph—a tank here, an armored vehicle there, and foxholes with bored English privates sitting under netting or canvas. Atop a hill near the Iraqi border the cavalcade skidded to a dusty halt. The Inniskilling Fusiliers, a bereted, brash Scots outfit, held the line, but reporters took the hill with ease. Regimental officers were engulfed by correspondents. Pencils scribbled in answer to innocuous questions. Cameras clicked.

"The Iraqi Army is building up over there beyond the border," an officer said. He stretched a hand toward the horizon, which merged into heat waves lifting from the desert.

One reporter caused an awkward moment when he asked the officer how he knew. The reporter earlier had been expressing his pacifist views in a loud voice, as he wondered what British troops were doing here. And he had asked his question with an undercurrent of hostility, as if he thought this officer might have been responsible for the whole ridiculous situation of two opposing armies facing each other.

But this tall, blond Scotsman with a reddish fuzz on his chin wasn't one to be overwhelmed by any reporter. Coolly, he stared at his inquisitor and answered: "I can smell their shit."

Reporters, who had pencils poised for a description of some new detection device, snickered. Everybody squinted "over there," but nobody saw anything, not even the movie cameramen who have unusually sharp eyes for the unusual, the odd, or the violent. The temperature was 115 degrees.

British correspondents, who predominated, hurried among the men in foxholes. Notebooks were filled up with "color" for the folks back home, concerning the doings of Private Bert Allen, Private Phil Jones, and various other privates who seemed to be doing nothing.

A horn honked. In this heat, it required little to lure correspondents to the air-conditioned taxis. The motorcade resumed its roving. Now individual initiative began to assert itself. A photographer in one car would spot a potential picture. His taxi would veer from the herd, race to the subject.

Photographers in other cars would sense that they were being scooped. More cars would veer. Then the whole pack would descend on the subject. That might be a mess tent, a machinegun post, or a water-tank truck. Any water was popular. The temperature was about 120 degrees.

One enterprising photographer located two Kuwaiti privates occupying a pup tent amid sand dunes. Obligingly, privates posed while crawling along the sand, rifles carried in folded arms, fierce scowls on faces. One could imagine the caption on the picture: "Two Kuwaiti soldiers on a scouting mission."

Other photographers sensed the photogenic possibilities. A swarm of them descended on the scene, surrounding the models.

"Hey! You're getting into my picture," pleaded Photographer Number One. "Why don't you find your own soldiers?"

Nobody paid any attention. News photographers seldom worry about anyone else's feelings. The temperature was about 121 degrees. Emotional temperature of the first photographer about the same.

A horn honked once more. Correspondents and photographers scrambled into cars. A hot wind blew harder, raising gusts of sand. The temperature crept up to 125 degrees in the shade, with no shade in sight. The party headed for home, and I thought of a remark made by one British army officer: "The shooting could start tomorrow. This is a tense situation when you have two armies in a faceoff."

But it did not start. The firm British answer to the threat from Iraq forced a change of intention in Baghdad. That crisis, like others in the area at different times, receded to that backburner where dozens of simmering pots now sit.

Looking back, this particular crisis may seem like comic opera.

Actually, the moral is far more significant. For a hundred and fifty years, whenever a crisis brewed, the British appeared on the scene to solve it. They provided the backbone for any stability to be found in the Gulf. They mediated between sheikhs in tribal wars, deflated territorial ambitions of desert lords, and trained local soldiers in policing duties.

Then, early in 1968, Britain's Prime Minister Harold Wilson made a major blunder. Without consulting any of the Gulf nations and sheikhdoms, he unilaterally announced that Britain would withdraw forces from the Gulf by the end of 1971. The richest petroleum area in the world was to lose its policemen, not at the behest of Gulf peoples but because of a London decision allegedly made to help the pound sterling. Britain felt it could not afford to play policeman any more.

Immediately, there was an upsurge in subversive activities by groups seeking to radicalize governments of the Gulf. There were fears that territorial claims might lead to disruptive wars. Kuwait, now independent, anxiously wondered if the Iraqi threat would rise again. It did, of course, in late March 1973, when Iraqi troops overran a Kuwaiti border post, and loudly voiced claims to parts of Kuwait. By July 1973, Kuwaiti and Iraqis were negotiating their problems across a table. But Kuwait was ordering more weaponry with some of its oil profits.

Meanwhile, the Soviet Union had achieved close ties with Iraq. Iran responded by increasing the effectiveness of its defense forces. British diplomats promoted a federation plan that had been around awhile without generating much enthusiasm: the linking of seven sheikhdoms of the Trucial States plus Qatar and Bahrain into a federation. Stability might be fostered and the oil would keep flowing to refineries of the world.

Qatar and Bahrain opted for their own separate independence, and on December 2, 1971, the seven Trucial States agreed to federate as the United Arab Emirates, the day after Britain terminated all protectorate treaties. The states are Abu Dhabi, Dubai, Sharjah, Ras al-Khaimah, Umm al-Qaiwain, Ajman, and Fujairah. All are strung along the lower end of the Persian Gulf, except for Fujairah, which is on the Gulf of Oman.

The first three have been described. Ras al-Khaimah is a small enclave with a capital pressed between mountains and the sea. Scenically, it is the most beautiful of all the sheikhdoms, with a modern hotel boasting a swimming pool and a gambling casino.

That casino, the only one in the Gulf, is virtually empty for six days a week. Every Thursday night, the Moslem equivalent of Saturday night, it swarms with business and oil men from nearby Sharjah or Dubai.

You pass through both Ajman and Umm al-Qaiwain to reach Ras al-Khaimah from Dubai. Ajman is only a few miles wide and you might miss it if you bent your head to light a cigaret. You would not miss much—a scattered collection of huts beside a small port that cannot accommodate much more than rowboats. Ajman's ruler, Sheikh Rashid bin Humaid al Naimi, is a benign old gentleman with a white Santa Claus beard, flowing robes, and gold slippers. He has ruled since 1928, which makes him the longest-reigning sheikh in the Gulf.

His neighbor, Sheikh Ahmed bin Rashid al-Mulla of Umm al-Qaiwain has ruled almost as long, since 1929. He, too, has a white beard, but it is more like a well-trimmed goatee. For years, he has been the area's peacemaker, the arbitrator of tribal disputes. Today, in ailing health, he no longer entertains visitors by bending a silver dollar between his powerful fingers.

Sheikh Saqr bin Mohammed al-Kasimi, fifty-six, ruler of Ras al-Khaimah, is a shrewd, hard bargainer who caused the most trouble during United Arab Emirates negotiations. He hoped that an oil strike would give him more bargaining power when jobs of the new federation were allocated, but this did not happen. He entered the union as leader of one of its poorest countries. Fujairah is just as poor, a backwater reachable only by four-wheel-driven vehicles on the land approach from neighbors in the federation.

The first president of the United Arab Emirates is Sheikh Zaid of Abu Dhabi. Sheikh Rashid of Dubai is the vice-president. The sheikhdoms retain autonomy while the UAE manages security and external affairs. The member states seek to coordinate their financial and development policies to avoid duplications and spending inefficiencies.

It is hoped that the UAE will promote stability in an area that easily could divide into bickering sheikhdoms too weak to oppose disrupting influences from outside, such as the Chinese-supported Front for the Liberation of the Occupied Gulf, which provides arms and a Marxist ideology to rebels in the Dhofar Province of Oman. The organization has broader aims, too, as literature dispensed in the Gulf indicates. It seeks radical, far-left governments in all of the states of the Gulf.

The Front is only one of several subversive groups active in the area. There are also Iraqi agents who peddle anti-sheikh propaganda and radical Palestinian groups who collect funds, use threats against Palestinian workers in oil fields to extort additional money, and stir up trouble in other ways.

X

The Iraqis

Over a hundred passengers from the Middle East Airlines flight waited in line in the air terminal at Baghdad for immigration checks. Without air-conditioning, people in the building sweltered. Passengers crowded together in the drab lounge, trying to improve positions in the queue, adding to the discomfort. It was apparent that immigration officials meant to check closely each arriving passenger and his passport.

Three perspiring Russian oil technicians in rumpled suits stood in the line just ahead of me, three abreast as were most of the people in the line by this time. A steward had told me that Russians were destined for the North Rumaila field that Iraq developed with assistance of the Soviet Union.

A Kuwaiti in white robes fanned himself with his passport. Some banknotes protruded conspicuously from its folded pages. If meant as a bribe for service, it was a futile gesture. He waited with the rest of us. So did the Russians. But not for long.

One of the immigration clerks glanced up. There is something about the Slavic features and blond hair of the Russian which makes him easy to spot in an Arab crowd, especially when at least half of the Arabs are in robes and keffiyehs. The clerk motioned for the Russians to come to the head of the line.

"We go Rumaila," one of the three said in English.

In minutes they cleared immigration and were whisked away. It was only a small thing, but it was a reminder that Russians are favored in Iraq today. Ever since Egypt's President Gamal Abdel Nasser closed an arms deal with Czechoslovakia in September 1955, the Communists have had a toehold in the Middle East.

Relations between Egypt and the U.S.S.R. cooled in 1972 when Egypt's President Sadat dismissed his Soviet military advisors. They warmed again after the Yom Kippur War when Russia pressed for a cease fire favorable to Egypt. Iraq has provided another and perhaps more important base for the U.S.S.R. in the Middle East, because unlike Egypt, it has footage on the Persian Gulf. And the Persian Gulf is the Western world's key source for oil.

It is ironic that Iraq, which used to be the most anticolonialist of Arab states, now is not only a client of the U.S.S.R., it even concluded a fifteen-year treaty of friendship with it in March 1972. An aim of the 1958 revolution in Iraq ostensibly was to shatter political ties with powers outside the Middle East, but now Iraq has tied itself to the U.S.S.R. Yet the Iraqis in government offer all kinds of reasons why this is natural.

The Iraqi has always been unpredictable because there is no composite Iraqi. Ethnically, the country is a racial and religious mosaic, and that mosaic is disorganized, with little cement to bind it together.

One Arab proverb says, "The Yemen is the cradle of the Arab race. Iraq is its grave." This is a bleak assessment. Yet Iraq often has been a troubled land. Again part of that is due to the diversity of the nationalities and religions to be found here. A California-sized nation, Iraq has a population of about ten million, composed of a patchwork of groups that feel little affinity. In the marshes of the south, where the Tigris and the Euphrates form a joint delta of lakes and channels, live the Maadans, a marsh-dwelling people descended from the Sumerians, Babylonians, Persians, and Arabs. In the north are the Kurds, perhaps as many as 1.5 million, who claim descent from the ancient Medes and speak an Indo-European language. Numerous Iranians reside in the country, a minority that now is persecuted because of the political troubles between Iran and Iraq.

There are minorities of Armenians and Turkomans, too. The religion is predominantly Moslem, yet there are Nestorian, Chaldean, and Assyrian Christian minorities. The Moslems are divided into Sunni and Shiia on a 45–55 percent basis, respectively. The

Shiia are farmers and village dwellers, often poorly educated, often ripe for exploitation. The Sunni are mainly urban dwellers, usually better educated than the Shiia, often found in government posts, sometimes in positions to exploit the Shiia, or Shiites.

One Western diplomat with long service in the country sat at a table in the Alwiyeh Club recently and dissected the political and social structure of Iraq for me. This Baghdad establishment once was a colonial stronghold much frequented by British oil men, military officers from the United Kingdom army on training missions to Iraq, and diplomats who seemed to have little else to do but sit at terrace tables at the club. Long after the 1958 revolution ended Iraq's neocolonial ties with Britain, the club retained its atmosphere.

"Iraq is not a country in the Western sense of the word," my diplomatic friend said. "It is a hodgepodge of nationalities and religions which have been tossed together by fate and history. It can only be ruled through a strong central government in Baghdad, one which is not afraid to be ruthless in seeking its ends."

The artificiality of Iraq's creation is revealed in words of Winston Churchill. In 1921, he declared in the House of Commons in London that it was British policy "to attempt to build up around the ancient city of Baghdad, in a form friendly to Britain and her allies, an Arab state which can revive and embody the old culture and glories of the Arab race."

In this cool and deliberate manner, this part of the dismembered Turkish Empire came into being as a national state. In 1921, the British installed Faisal ibn Husain, a descendant of Mohammed, as King of Iraq. From the first, the country had a stormy existence, owing to its divergent ethnic and religious currents. The royal family clung to the throne, thanks to the British backbone injected into the army through training missions and equipment. But prime ministers and their governments came and went so fast that historical accounts of the 1921–58 period read like a biographical index.

In that period, there were fifty-eight different governments in power. Had it not been for the discovery of oil in the Kirkuk area in 1927, Iraq might have remained an undeveloped backwater, torn by tribal feuds. But oil provided the revenues for an infrastructure, and also opened the country to Western influences. With education came a rise of nationalism. Often this has seemed to be more of a negative than a positive force, an influence first against the British, then against the West after the 1958 revolution.

"It is easier to hate than to love," said Fouad Khorney, a young irrigation engineer encountered in Baghdad. "We had to win control of our country before we could progress. So we had to be against those forces which wanted us to remain under colonialism."

"But even today you seem overly suspicious," I said.

"We have reason to be because of your support for Israel," he answered. "It was America which first showed enmity by backing our enemy. We had no other choice than to turn to Russia."

Iraqi troops took a bloody beating in the Yom Kippur War, adding to the hate for Israel's supporters. In pique Iraq nationalized Exxon and Mobiloil holdings in IPC, last U.S. equity there.

One encounters a naivety among Iraqis where Russians are concerned, a belief that Iraqis are far shrewder than the Russians, that Iraq will always have full control of any partnerships that might be arranged with the Soviet Union. One also encounters a chip-on-shoulder brashness about Iraqi's nationalism. The typical citizen has a fierce pride that is ready to take offense at the least slight, real or imaginary. In the souks and back lanes of Baghdad, a tourist with a camera is apt to encounter some belligerent self-appointed "policeman," usually a youth in his teens, who will warn against taking any pictures of ordinary life on the streets or of ancient buildings. New buildings are all right, especially if they happen to be schools or hospitals. But proud young Iraqis want no pictures that show the poverty and distress of some of their slum areas. It never seems to occur to them that if they devoted more of their energies to eliminating the slums, the ghettos wouldn't be there.

Iraq's nationalism exploded on July 14, 1958, when units of the army revolted. King Faisal II and the crown prince were slaughtered, and power was seized by Brigadier General Abdul Karim Kassem, a moody, neurotic officer who was to live out his life in mortal dread of another coup. He took an anti-Western stand immediately. The American government feared that he represented the takeover of the whole Middle East by nationalistic, anti-Western Arabs who probably would affiliate with Egypt's Nasser under a Russian umbrella. President Eisenhower ordered protective troop landings in Lebanon, a little democratic friend of the West.

American troops did not remain long in Lebanon. Meanwhile, Kassem consolidated his power. In June–July 1961 he sought unsuccessfully to annex Kuwait. In that same year, he nationalized undeveloped oil concessions of Iraq Petroleum Co., Ltd., the consortium that had been producing oil since the initial discovery in the country, at Kirkuk in the north, and through its affiliate, Basrah Petroleum Company, in the south.

A key property in the nationalization was the North Rumaila concession, adjacent to a rich property of Basrah Petroleum. This turned out to be just as rich as expected, and Iraq obtained Russian help to develop the field.

While this nationalization was a significant development, an even more important occurrence took place in 1960. This was the formation of the Organization of Petroleum Exporting Countries, or OPEC. Roots of this organization go back to the nationalization of Iran's oil industry in 1951. Iran's subsequent troubles stemming from a lack of technicians showed how helpless a lone nation might be when facing the oil industry. But what if producing countries supported any nation caught in a dispute with the companies? Would not this force companies into line? Questions such as these lay behind the formation of OPEC.

Ideas sometimes lie dormant awhile in the Middle East, until some emotional issue heats the atmosphere enough to draw the ideas into the sun. In 1959, oil companies provided the issue by slashing posted oil prices because of a glut of oil on the market at the time. Since tax revenues of nations are based on posted prices, this meant that government revenues might be reduced too.

Immediately, Arab leaders protested. It seemed incredible to them that decisions of oil companies could force the Arab nations to reduce budgets. Agitation for a producers' "union" flared.

Perhaps it was appropriate that Baghdad was chosen as the site for the meeting in September 1960 that was to bring OPEC into being. It certainly was significant that Iraq submitted the resolution calling for creation of an organization that would meet regularly with the object of "coordinating policies and attitudes in eventualities concerning oil production." Founder members were Iran, Venezuela, Saudi Arabia, Kuwait, and Iraq. Mohammed Salman, then director of the Arab League's Petroleum Bureau, attended as an observer. The Saudi Arabian delegation was led by Abdullah Tariki, the strident nationalist who attended the University of Texas, then worked as an executive trainee with Texaco, Inc., in western Texas and in California in the 1945–49 period. Even then he already had adopted a slogan: "Arab oil belongs to the Arabs." Already he was an impatient, emotional, anti-Western force who was to play an important role in the formation of economically revolutionary ideas in such countries as Iraq, Syria, and Algeria. Already he was getting too revolutionary for his native land, though he was shortly to become his country's first oil minister, only to antagonize Faisal, then the crown prince. He departed from power by request, to establish himself as a petroleum consultant in Beirut, where he became one of the foremost proponents of oil nationalization in the Arab world.

The oil companies practically ignored the birth of OPEC. In Beirut, shortly after its creation, I listened to one executive of a Western oil company as we sat in the bar of the Saint Georges Hotel, the seaside hostelry that has always been popular with oil men and Arab sheikhs of the area.

"OPEC?" He could scarcely conceal his contempt. "That's just another one of these Arab talk forums," said he. "Like it or not the only place the Arabs can sell their oil is in the West. So they have to deal with us."

"On company terms?" I asked.

"On jointly negotiated terms," he said. "We try to be fair, and I think we have been."

I asked him what he thought of Tariki, who happened to be occupying a suite at the Saint Georges Hotel at the same time. In fact, that same morning I had sat with Tariki and listened to him outline how producer countries should be sharing in profits of oil right down to the corner filling stations in America and in Western Europe.

"Tariki?" The oil executive showed his contempt. "He can't even get his facts straight." He laughed as he explained how a paper presented by Tariki at that year's Arab Oil Congress had contained an error of two billion dollars in the calculations.

Arabs, generally, aren't adept at handling details with exactitude.

Sometimes the substance of a presentation may not seem as important as the manner in which it is presented. If one's arguments are resounding and loud, then listeners should not question the details. Perhaps oil-company representatives were paying close attention to details, while the Arabs were hearing only the broad outlines of Tariki's anticompany arguments.

Yet OPEC, initially, did seem to be only a talk forum for presentation of resounding debating points. It was in June 1961 that Mohammed Salman, then Iraq's Minister for Oil, provided some clues about OPEC's aims and intentions. I met him in an office of the ministry where a window air-conditioner was holding back the 110-degree Fahrenheit temperature outside.

There was the usual tea, the usual introductory pleasantries that customarily precede any discussion in Arab lands. Then he asked a rhetorical question when the conversation settled upon OPEC. "What," asked he, "do you think would happen if all the oil-producing countries stood together as a bloc in negotiating with oil companies?"

"Oil production might stop," I said. "Maybe your revenues would stop, too, as they did when Iran nationalized."

Salman shook his head. "Up to now the oil companies negotiated with us one at a time. We are bound to lose if we continue in this fashion." Leaning forward, he placed emphasis on his words. "Up to now companies have not considered producing countries as partners. We intend to be treated as partners."

He rose, strolled to a window, and turned down an air-conditioner that was blowing a pleasant breeze into the spacious office. I would have liked more frigidity from the air-conditioner rather than less. He returned to his desk to outline his dreams for OPEC, fingers playing all the while with a pin cushion filled with the pins used as paper clips by Iraqi bureaucrats.

Within a decade OPEC certainly did live up to his most glorious dreams for it. Qatar, Libya, Indonesia, Abu Dhabi, Algeria, Nigeria, Trinidad and Tobago, and Egypt eventually joined. By 1973, it could claim to represent 80 percent of world oil exports. Recently, in a burst of enthusiasm, one OPEC official declared, "What OPEC wants, OPEC gets." The boast seemed appropriate.

Iraq, however, had its own troubles long before OPEC reached the height of its glory. Many of those troubles were provided by the Kurds, a race that resents the dominance of the Arabs in Iraq.

While Kurds do not like the Arabs, the latter despise the Kurds. In troubled times, which is apt to be most times in northern Iraq, nearly every Kurdish male will tote a rifle, perhaps with a bandolier of bullets across his chest or around his middle. Moreover, Kurds know how to shoot. Where Arabs are concerned they are apt to shoot first and ask questions later.

An old Arab proverb in Iraq says, "There are three plagues in the world: locusts, rats, and Kurds."

In Baghdad, one young Iraqi, who was completing an engineering course at Baghdad University, showed his concern when I mentioned that I would be traveling through Kurdish country in the mountains of the north. He warned, "Never trust a Kurd. That is a good way to get your throat cut."

He made a gesture with his finger along his throat.

Later, near As Sulaymaniyah, a Kurdish city of nearly 100,000 in the north, I repeated this story to a Kurdish coffee-shop proprietor, a big, red-bearded Kurd with a tarboosh on his head and baggy trousers that might have made a sail for a small yacht had he unraveled a few seams.

"The Arabs are the curse of Iraq," he said. "They talk out of both sides of their mouths, sometimes at the same time."

Kurds revolted against the central government in 1927, 1931, 1932, 1933, 1935, and 1943. There were periodic troubles in the 1943–58 period. Then for the following twelve years, an almost continual war was waged by forces led by short, muscular Mullah Mustafa Barzani, Kurdish leader, against the Iraqi government. Kurds have not been able to establish their own autonomous state, which might have joined together Kurdish elements in Turkey and in Iran, too. Neither has the Iraqi government been able to establish its authority in the northern part of Iraq.

The Kassem government, which seized power in 1958, was plagued by Kurdish rebellions through most of its existence. Every summer there was a major offensive by Kassem's army, and key areas in the north would be occupied. Then through the fall and winter, Kurdish guerrilla raids would force withdrawal of government troops, and the whole pattern would begin again. The war bled Kassem's regime, contributed to the discontent that always seems present in Iraq.

When a military coup in February 1963 ended Kassem's dictatorial rule, he was seized and shot. Mobs roamed the streets of Baghdad and many an old personal score was settled. In any change of power, Iraq is not the safest place in the Middle East.

The Baath Party was behind this particular coup, with Colonel Abdul Salem Mohammed Arif and Brigadier General Ahmed Hassan al-Bakr leading it. The Baath Party is a violently nationalist, left-wing group that views all Arabs as belonging to one "Arab nation." It is socialist in political orientation, advocating that all Arabs are equal and that class distinctions should be abolished. Arab unity is a key plank in its program. Thus, the Baath Party does not feel itself bound by country boundaries, though it has succeeded in taking root only in Iraq and in Syria.

Baath was created officially on April 6, 1946, in Syria through the efforts of Michael Aflak, an orthodox Christian Arab, and Salah Bitar, a Sunni Moslem. They preached that Arab failures are due to the decadence of the Arab spirit. By re-creating an Arab consciousness, the Arab nation could be reborn. Ultimately, there then would be an Arab nation stretching from the Atlantic to the Gulf.

"The Baath is more than a party. It is a state of mind, an atmosphere, a faith, a doctrine, a culture, a civilization with its own intrinsic worth," says Salah Bitar, sixty-two, the Sorbonne-educated, one-time teacher, long-time politician who now spends most of his time in Beirut in exile. He is a mournful-looking intellectual with thinning, gray hair, and heavy bags under eyes, as if he now might spend most of his time reading in dim light. That sadness may stem from the fact that Baath has not lived up to the dreams he had about Arab renaissance when the party was created. Today he feels that Baath aims have been perverted by the politicians who have expelled him from the party he founded.

Grandiose platforms always have been diluted by the particular leaders who happened to be heading the Baath Party in Syria and in Iraq at different times. In Iraq, the 1963 revolution ushered in an internecine power struggle among various wings of the party and between the party itself and non-Baath, pro-Nasser elements. No matter who happened to be in power, however, the government had to prove its nationalism by being anti-IPC, by averring that the oil is a "national heritage," and by regarding Britain and America as colonial powers with anti-Arab intentions. This led almost naturally to an ever-closer rapport with the Soviet Union.

For all the talk of "freedom" and "rights of the people" that one hears from Iraqi politicians, the truth is the governments are unfamiliar with Western-style freedom of the press. Even a Western-style press conference may be unsettling to Middle East politicians, who are accustomed to handling their activities behind shields that are seldom pierced by the average man-in-the-street.

I happened to be in Beirut on the day of the February 1963 coup. Beirut is always a good place to be when trouble flares in the Middle East. This is one city in which you do have a free and active press. So Beirut has become a listening post for following all of the intrigues and conspiracies of the Middle East. It has a pleasant cocktail circuit, too, a comfort rather than a hardship post for the individual who must do the listening.

I joined a group of other newsmen in chartering a Vickers Viscount for an unscheduled flight from Beirut to Baghdad. The moment shooting starts anywhere in the Middle East, regular airlines cancel flights. No airline likes to risk its high-priced planes, or passengers who might have lawsuit-prone relatives. Fortunately, charters may be arranged, at a price, sometimes with an appropriate signing of waivers.

The flight was uneventful, even though the civilian-control tower at the Baghdad airport was unoccupied, with all operations in the hands of the military at the other end of the field. There were about seventy newsmen in the party, most of them American, with a scattering of English, French, and German correspondents. The party was escorted to the Baghdad Hotel in the town's center and a request for a meeting with new rulers of Iraq was quickly granted. Those new rulers wanted to assure the world that the people's will had triumphed.

Whenever a strong man or a group of strong men seize power, the new boss never publicly says, "I just wanted to run things myself, so I took over." No. The usual practice is to sanctimoniously claim that "the people" are behind the coup. Overnight a whole propaganda facade may be woven by adroit revolutionary hangers-on who are adept at repeating such words as "imperialism," "neocolonialism," and "exploitation" often enough to make them sound valid, at least to themselves. Few people are better at believing their own propaganda than the Arabs.

This particular press conference began with a "cocktail party" in the Baghdad Hotel's mahogany-paneled private salon. About thirty of Baghdad's reportorial talent had emerged from whatever shelters they had been occupying through the bloody hours of the coup. They were the only ones who appeared in the salon. "Cocktails" in this predominantly Moslem country consist of Pepsi-Cola or tomato juice.

The hotel's nearby bar was operated by bartenders who were non-Moslem in spirit if not in fact. There, foreign reporters congregated over Scotches and gins while government aides futilely wailed that the press conference really was next door. Then President Arif and Prime Minister Al-Bakr swept into the hotel's plush lobby, accompanied by about thirty soldiers lugging cocked burp guns. Reporters and photographers poured from the bar, emptying glasses on the run.

A beefy captain in the lead, with a machine pistol under his right arm, raised his left hand like a traffic cop. Obviously, his authoritative air, uniform, and gun always had drawn respect in the past. This evening, the press stampede overwhelmed him.

"Allah," he cried as he disappeared in the melee. Soldiers leveled their weapons uneasily. Reporters brushed the barrels aside and surged toward the country's new leaders. Flashbulbs popped. Newsreel cameramen pointed lenses at the approaching dignitaries, walking backward before them. Soldiers formed a tight, moving mass about the leaders, waiting for a signal to open fire. None of them had slept for three days. One pop of a champagne cork would have started a massacre.

Smiling woodenly, President Arif, neatly dressed in a double-breasted suit, and Prime Minister Al-Bakr, in a brigadier's uniform, forced their way through the press mob into the hotel's private salon, which no longer was very private. Half walking, half pushed by reporters, they surged toward a corner divan.

Ray Maloney, a UPI Middle East man, raced to the divan and

plunked himself on an arm. The crush wedged him tightly against the prime minister. George de Carvalho, a *Time* magazine man, grabbed a chair directly in front of the divan. As other correspondents pressed forward, he was pushed almost onto President Arif's lap.

Some correspondents wriggled by the befuddled armed guard. The soldiers linked arms and fended off other journalists. Those in the select inner circle fired questions through an interpreter. Nobody more than three feet away heard a thing. "This is an outrage. What kind of press conference is this?" yelled George Weller, the *Chicago Daily News* man from a position far in the rear.

"Down in front!" cried one photographer, trying to lift his camera above the sea of heads. One novice reporter obligingly dropped to his knees and was pinned so tightly by the mob that he helplessly retained that position through most of the interview.

Every extra table and chair became a photographer's perch. The furniture was of a style that might be called Iraqi Swedish, definitely not made for height building. One cameraman was upright on a chair when the cushion collapsed. He found himself standing on the floor with the chair around his knees.

"We can't hear a thing! This is an outrage!" cried Weller from amid the pack. Maloney was scribbling furiously on a pad as the interpreter relayed answers in conspiratorial tones.

"What did he say? What did he say?" a dozen voices kept repeating.

"He says the people now are masters of their destiny," repeated a government official in the crowd.

"Don't push! Don't push!" shouted Musar al-Rawi, the slender, scholarly Minister of Guidance, a bespectacled man encountering his first American-style press conference and not relishing it at all.

The President and Prime Minister rose. Soldiers drove a wedge through the mob. There was a shatter of glass and the crash of another chair. Photographers fought for final pictures as the official party departed. Two wire-service men huddled on the divan comparing notes and coming to agreement as to just what the President and the Prime Minister had said. Several dozen other reporters waited impatiently. They had been in the rear ranks and had not heard a thing. They wanted a second-hand version of the rehashed tale being prepared by the wire-service men.

The manager of the hotel entered, stared in anguish at the smashed furniture. When newsmen checked out of the hotel they found neat packets of two-day-old press handouts in their key boxes, along with stiff hotel bills padded by 25 percent for broken furniture.

Iraq had few such press conferences. From 1958 to the end of 1970 the country had press, mail, and cable censorship. Prime Minister Al-Bakr fell from power in the internecine squabbling, only to return to power in another coup d'état on July 17, 1968. Sometimes this particular government seemed to be suffering from paranoia as it ruthlessly eliminated potential opponents and "Israeli spies." In 1969, there were over fifty political executions. Some were staged in a central square in Baghdad as warnings to the general public. After an abortive coup in January 1970, another score of alleged plotters was executed.

Relations with the United States were severed following the June 1967 Israeli-Arab war. As mentioned, minor American oil interests were nationalized after the Yom Kippur affair. Yet, the visiting American may encounter friendliness as long as he doesn't flaunt his nationality.

"We don't have anything against the ordinary American. It is your government policy we do not like," we were assured by one newspaperman friend. I had hesitated to contact him, fearing that we might get a rude welcome in the light of the political climate. Moreover, I feared that, were we welcomed, we might compromise a friend in the eyes of the government. Our friend had heard somehow that we were in Baghdad, and had called on us. He seemed unconcerned at our anxieties about his being seen with us.

"The Ministry of Information knows you are here," he said, philosophically. "They know you are quite harmless."

Yet, even as we were talking about mutual friends and of past events jointly shared, I could read the headline on a copy of the *Baghdad Observer* that I had picked up. "Communist's Participation in Cabinet Would Help Foil Imperialist Schemes," it said. The story detailed how a couple of Communists had been named to the cabinet; then it reprinted a Communist Party statement.

"The imperialists and reactionaries have been exerting their endeavors to prevent any cooperation between national parties in our country," it said. It went on to say that entrance of Communists into

Iraq's government "will be viewed with displeasure in imperialistic, Zionist and reactionary quarters."

Internally, Communists never were much of a factor in politics here, or anywhere else in the Arab world, for that matter. Red parties operated with only a handful of adherents, often latching on to the handiest available radical party rather than seeking to battle alone. Externally, however, Iraq has been one of the most receptive nations in the Middle East to the blandishments and "friendship" overtures of Moscow.

Just what have the Russians been up to? They seem to have at least a trio of reasons for trying to win a place for themselves in the Middle East. In the first place, the Soviet Union undoubtedly would like to assure petroleum supplies to its client states in Eastern Europe, and perhaps to assure itself of an import source in the 1980s. In the second place, it realizes that the whole industrial non-Communist world, with the exception of Canada, now faces an energy gap, and the Middle East is the key supply source for closing that gap. Therefore, if the Russians can exert pressure in some ways on that flow to the West it will have its fingers on what may prove to be the Achilles' heel of the Western world. Thirdly, Russians probably want all the diplomatic mileage they can get in an area that has been notoriously unstable politically for decades. Governments are sure to change again and again in several states of the Middle East in the next few years, if only because aged kings and sheikhs will be passing on. It is in the U.S.S.R.'s interests to have more radical governments appearing in those countries. It is not averse to helping bring about that result through subversion when opportunities arise.

It certainly is in the U.S.S.R.'s interests, too, if the world price of oil rises up and up into the stratosphere. The Soviet Union is virtually self-sufficient in oil. It exports on average about a million barrels of petroleum daily to non-Communist countries. The higher the world price, the more foreign exchange the U.S.S.R. will be earning from its oil exports.

Moreover, the higher the world price, the more the United States and Western Europe will be spending of their foreign exchange. Thus, Russians have positive reasons for helping the Arabs to obtain whatever price the oil market will bear. It would be difficult, if not impossible, for Russians to bar that Middle East oil to the West, for the Arabs have to sell most of it to the West and Japan. But Russians have every reason for wanting to see that the West and Japan pay dearly for their oil.

The Soviet Union now is the only major power that is producing oil in excess of its own needs. This certainly is an enviable position at a time when the United States faces possible energy shortages. The situation is a cause for concern to every American and Western European. A nation or group of nations that face energy shortages will have more difficulty maintaining industrial and technological leadership over a nation that may have an abundance of energy sources.

In 1973, however, a small cloud was appearing on the Soviet energy horizon, too. In an article in the Soviet journal *Neftyanoye Khozyaistvo*, V. D. Shashin, U.S.S.R. Minister for the Oil Industry, admitted that the growth of oil reserves in the U.S.S.R. is lagging behind the growth of production. Enormous natural-gas reserves have been found. But these are proving difficult to exploit.

Are the Russians seeking oil sources in the Middle East because they realize that they, too, may have energy troubles in the 1980s? The auto age is coming to the Soviet Union, and oil consumption is rising rapidly. Anyone who visits Moscow in the atmosphere of detente the Russians are now creating finds that not only are Soviets interested in Middle East oil, they also are interested in United States technology for development of oil and gas fields within the Soviet Union.

"Yes. We are interested in production and service contracts with Western operating oil companies," Victor I. Mishchevich, U.S.S.R. Deputy Minister of Petroleum, told me. I sat across a table from him in the Ministry's building along the Moscow River embankment in the Soviet capital, as he related how his ministry already has had discussions with British Petroleum, Mobiloil, Union Oil, Occidental Petroleum, and other companies.

It was a curious meeting, for with only ten minutes' notice I had been whisked by a government official from my room in the Intourist Hotel to the Ministry on a hot August day. "The Petroleum Ministry wants to see you," I had been told.

At the Ministry I was led into a conference room where I found myself a participant in a negotiating round between a half dozen top officials of the Ministry and three executives of Dresser Industries, Houston, Texas, which was considering possible joint production of oil drilling equipment with the Russians.

"We have nothing to hide," Deputy Minister Mishchevich assured me. But I couldn't help but reflect that it is not normal for a government ministry to open negotiations of this kind to a newspaperman. It was evident that the Russians had a purpose in so doing. They wanted to publicize how welcome American oil companies now are in the Soviet Union. They would like to win American help at the same time as they gain a better foothold in the Middle East.

U.S.S.R. crude-oil production has been mounting steadily in recent years. The following are statistics of attained and projected output for the Soviet Union:

	Barrels per day
1960	2.96 million
1970	7.1 million
1971	7.5 million
1972	7.9 million
1973	8.5 million
1975	9.9 million

These figures do show that the growth rate in the U.S.S.R. has slowed in the 1970s from the 1960s. *Oil and Gas Journal* estimates Soviet reserves at about 75 billion barrels in 1972. American reserves were estimated at 36.5 billion barrels by the American Petroleum Institute, or half those of the U.S.S.R.

Mishchevich refused to confirm the *Oil and Gas Journal* reserve estimates, which amount to a 24-year supply at present production rates. But when I mentioned a much lower estimate of only a 13-year supply made by a British source, he shook his head: "We have much more than that," he said.

Admittedly, the Soviet Union has a vast potential for the development of new oil fields in Siberia, and it already has enough natural gas in proven reserves to meet ninety years of consumption at the 1972 pace. But many of the potential fields are in remote areas, far from points of consumption, with staggering amounts of capital needed for their development.

Meanwhile, Soviet consumption is rising, as is that of its Eastern European satellites. The present five-year plan ending in 1975 calls for the shipment from the U.S.S.R. of fifty million tons of oil per year to Czechoslovakia, Hungary, East Germany, and Poland. Shipments in 1970 amounted to about thirty-three million tons, which indicates that current shipments are running an average of 51.5 percent higher.

"It may be entirely possible that Eastern European demands for oil will exceed Soviet capabilities to satisfy them by 100 million tons or more by the 1980s," says one research document of Radio Liberty, the Munich-based radio station that not only broadcasts to the Soviet Union but also maintains one of the world's best research agencies concerning activities within the Soviet bloc.

Thus, it is evident that the Soviet bloc as a whole probably will need substantial quantities of Middle East oil in the 1980s. When this is recognized, it becomes very clear why the U.S.S.R. has such a strong interest in the Middle East. And Russians will be very happy if American companies help them at home, for this would be a double guarantee of supplies for the U.S.S.R. American companies would get their payoff in oil and gas once new wells in Siberia started to flow, say Russians.

Iraq has been a willing ally for the Russians. When an impatient President Sadat ordered Russians to depart from Egypt in the summer of 1972, the U.S.S.R. already had in hand its fifteen-year treaty of friendship with Iraq. This rapport has its roots in oil as does much of the politics in this area. Ever since the 1958 Iraqi revolution, the Iraqi government and Iraq Petroleum Company, the consortium of Western companies, have been involved in a running battle over petroleum. In 1961, after Iraq nationalized company concessions that were not being exploited, IPC served noticed that it would fight in world courts should any oil be pumped from the concessions. That threat prompted Western oil companies to avoid any connection with the disputed area.

The Soviet Union came to Iraq's rescue, by offering credit, equipment, and manpower to develop the North Rumaila Field, a key property in the IPC-Iraq dispute.

"Thanks to the Soviet Union we have the opportunity to produce oil independently, to make an old dream come true," Adib al-Jadir, then president of Iraq National Oil Company, the government-owned firm, said. Another $70 million credit-equipment package in 1969 helped Iraq bring North Rumaila into production. First oil flowed in April 1972, with operations geared for a rate of 100,000 barrels a day. Immediately work commenced on a second stage to raise output to 350,000 barrels a day. Most of the initial output is going to the Communist bloc to pay for credits extended. Thus, Iraq evaded IPC's threats of court suits.

In the spring of 1971, IPC had a productive capacity of 1.7 million barrels a day, about 1.1 million barrels of that from IPC's three fields around Kirkuk in the north, and about 600,000 b/d from two fields of its affiliate, Basrah Petroleum Company, in the south. In addition, there was an insignificant productive capacity at another affiliate, Mosul Petroleum Company. Production in the north is carried via a 525-mile-long pipeline from Kirkuk across Syria to the port of Banias, with a second outlet extending to Tripoli, Lebanon. Shipments from Basrah Petroleum's fields are piped to the Gulf to a deep-water terminal at Fao.

Syria, a nonentity in the Middle East insofar as oil production is concerned, has a hand in the area's oil because of the Iraq pipeline and because of the Trans Arabian Pipe Line (Tapline), which crosses part of its territory to connect Saudi Arabian oil fields with Sidon, Lebanon. The latter pipeline has a 500,000 b/d capacity.

Syria charges oil companies for the right to operate pipelines across its territory, and believes in charging everything the traffic will bear. In 1972, the combination of Syrian excessive toll charges plus Iraqi profit squeezes finally created a situation toward which Iraq had been moving for years—its oil could no longer be priced economically in Western markets in competition with oil from elsewhere. So IPC reduced its production in Kirkuk.

Immediately, Iraq protested. Cutbacks hit the government where it hurt most, in the pocketbook. After weeks of arguing back and forth, Iraq, on June 1, 1972, nationalized Iraq Petroleum Company's Kirkuk fields with their 1.1 million-barrels-daily productive capacity. Curiously, however, Iraq left Basrah Petroleum alone, a wise gesture perhaps, since revenues from Basrah continued to flow into the government's treasury.

The Arab Oil Congress was in session in Algiers on the day of nationalization, an event that attracted oil ministers and executives from throughout the area. In attendance and on the speaking program was Abdullah Tariki, the doyen of nationalization in the Arab world. At 53, with streaks of gray in his curly black hair, he had lost none of his distrust of oil companies, nor had he lost any of his nationalistic fire.

News of the Iraqi nationalization flashed through the ornate conference hall before Tariki was scheduled to speak. He was in the back of the hall rather than on the platform when his turn came for a brief address. He rose to his feet to tumultuous applause when his name was called. Then, with the applause mounting, he strode down the aisle like a man who sees his life's dream becoming a reality. True, Algeria had already nationalized its oil, and Libya had taken over some operations. But radical Iraq was the first Arab Persian Gulf nation to seize control of a key field.

Appropriately enough, Tariki's paper at that meeting was entitled, "The Nationalization of the Arab Oil Industry."

Selling Iraqi oil was something else. IPC threatened action against anybody who bought it. Iraq's great good friend, the U.S.S.R., was immune to such threats but did not need a million barrels a day of Iraqi oil. Iraq's revenues dropped. It had started the year producing at a 1.9 million-barrels-a-day rate, finished with 660,000 b/d. The full 1972 average was 1 million b/d versus 1.7 million b/d in 1971.

The growing world energy problem saved Iraq. In early 1973, Oil Minister Sadoon Hammadi, astute, University of Wisconsin-educated economist, negotiated a settlement with IPC. Nationalizations were recognized, companies agreed to pay $345 million in back taxes and claims, and to help boost Iraqi output to three million b/d by 1975. As compensation, they received fifteen million tons of oil.

I saw Hammadi in his Baghdad office days before the Yom Kippur War, found him determined that Arab oil be a political weapon against Israel, happy that Iraq had "liberated" itself from company domination, and ready for much higher oil prices. In days, Arab oil producers raised prices by seventeen percent, then hiked posted prices, too, hitting companies with a total seventy percent increase.

France, meanwhile, provided a clue to the developing scramble for oil. Through its Compagnie Française des Pétroles, it agreed to purchase from Iraq 23.75 percent of the annual production of the nationalized Kirkuk fields over a ten-year period. Oil-industry sources fear that more bilateral, country-to-country deals may come, with nations bidding against each other to assure supplies.

In Amsterdam, at a Europe-American Conference in late March 1973, Walter J. Levy, a New York–based international petroleum consultant, warned of this development. Then he suggested that consumer nations should organize their own international group to coordinate oil policies. With America, Western Europe, and Japan forming a nucleus, such a group could be a countervailing power to OPEC. Consumer nations then might avoid the troubles that could appear if buyers start bidding against each other.

An oil buyers' cartel does not appeal to Arabs. Even a little nation like Lebanon, which has no oil of its own, believes in bargaining, especially when the odds are tilted in the Arab favor.

Lebanon has a laissez-faire philosophy that stresses market freedom, and it has some of the shrewdest merchants to be found anywhere. They are in the oil industry, too, though Lebanon does not produce a drop of petroleum. The country is in the business through its financial and trading know-how, factors that have made its capital, Beirut, the financial center for the Arab world.

In Lebanon, schoolchildren learn to count before they know the alphabet, and they do not bother with apples and oranges in their arithmetic problems. From the first grade on, they learn about money, and they probably already know something about foreign-exchange trading. As a money center, nearly everyone sees foreign currency every day, whether in tips or in deals negotiated with area businessmen. Money is bought and sold like potatoes, always with large commissions remaining on Lebanese fingers.

One story heard in Beirut provides some clues to the Lebanese character. A government educational inspector was visiting a school on an inspection tour. In a first-grade classroom, he asked that perennial question that sometimes establishes a rapport between a visiting adult and a bright tot: "How much is two and two, my boy?"

The boy's features screwed into a shrewd appraisal. He asked, "Are you buying or selling?"

XI

The Lebanese and Syrians

Nadia A. El-Khourey, a hazel-eyed, decisive blonde mother of five children, is an anomaly in the Arab world, where a woman's place usually is not only in the home but in a quiet corner of it. She is a businesswoman, a banker, and a director of Mothercat, Ltd., a holding company with twenty corporations under its control. Included is the biggest Arab construction outfit in the Middle East.

Sitting in a bright office in the Banque de l'Industrie et du Travail in Beirut, Lebanon, she smooths her hair with a womanly gesture as she says, "Let's see. I am the chairman of three companies, president of this bank, and on the boards of directors of six other companies. This is the limit under our new laws. I used to be on twenty boards but had to give up fourteen of them with this new law."

She says it without boasting, in the manner that a busy American clubwoman might list her clubs for a society editor. A photograph on a mantel shows her standing with a black-robed Roman Catholic cardinal who was passing through Beirut. Like many of the city's successful business people, she is a Christian Arab. Like just about all of them, she knows how to produce a profit. Her husband started the business. When he died, she assumed his role. Arab sheikhs regard her as "her husband's widow" rather than as an independent person in her own right. She is wise enough to overlook this as she bargains and bosses in executive style.

Everybody seems to be an entrepreneur in Lebanon. Even the taxi drivers are in business on the side, selling tours to Baalbek, offering selections of postcards, or hinting that there are certain women who might be available for a fee.

While neighboring Syria preaches socialism, Lebanon is a bastion of free enterprise. This country of 2.9 million people with an area smaller than that of Connecticut is the Middle East's supermarket and its counting house. The entire population sometimes seems out to make a profit.

The Lebanese see life as a business venture with eternal judgments weighed on the same scales as those used for buying and selling. You certainly cannot take it with you to the hereafter if you do not have it in the first place. Therefore, the Lebanese would just as soon have it in this life, for they know that, if any man does succeed in taking it with him, he probably will be Lebanese.

This person is unlikely to be Syrian, for socialism in Syria means sharing the poverty. Syria, with a population of seven million, has an area equivalent to North Dakota's, a state that it resembles in appearance. In earlier times the name applied to what is now Lebanon, Syria, Israel, and Jordan. Modern Syria was truncated by diplomats following World War I. It has suffered from unstable governments ever since—first through an unwanted French "protectorship," then through governments that have moved leftward—and the country stagnates.

Lebanon is half Christian, half Moslem, with a confessional government that seems precariously aligned, yet somehow seems to work. The president is always a Maronite Christian, the prime minister, a Sunni Moslem. Seats in parliament are divided on a ratio of six Christians to every five Moslems in the chamber. In all, ten sects are represented among the ninety-nine seats. The Christians are Church of Rome–affiliated Maronites, a sect that has survived for over 1,300 years in the Moslem world on the mountain range that traverses the country, farming their upland terraces and ambushing invading enemies. Even today, city-bred Lebanese speak nostalgically of life on "The Mountain," and they like to relax in the high country, perhaps on family picnics.

In Charlemagne's time, France negotiated the right to oversee Christian holy places in the domain of the legendary Harun el-Rashid. King Francis I renegotiated that right with the Turks, and the Maronites of Lebanon became special wards. In the 1860s, France even landed an expeditionary force to aid them. In the period between the two world wars, France held a protectorship over both Syria and Lebanon. Its rule of Syria was arrogant and ruthless; Christians of Lebanon enjoyed a benevolent rule that still leaves a residue of friendship toward France, at least among Christians of Lebanon. Many citizens speak French. Thanks to French missionaries and to Protestant missionaries from the United States, the Christians are the best educated of all groups in the country. They control much of Lebanon's wealth.

Syria, basically Moslem, has a mosaic of Armenian, Greek Orthodox, Syrian Orthodox, Nestorian, and Roman Catholic minorities. The Moslems are mainly Sunni, with Alawi, Druse, Shiia, and Ismaili minorities, and there are tribal differences. Urban dwellers disdain the rural populace, while provincial villagers distrust city cousins and live remote lives.

There is even a village, Maloula, which still speaks Aramaic, the ancient language spoken by Jesus Christ. It is thirty-six miles northeast of Damascus, reached by way of a rutted turnoff from the main Damascus-Homs road. At the end of a rocky canyon, mudbrick houses are perched on a steep hillside so perpendicular that some villagers use ladders between houses. The whitewashed church of St. Sarkis caps a rocky knob, its Byzantine cupola catching a late-afternoon sun.

"What does Aramaic sound like?" I inquired of Paulus, a bearded, black-robed Greek Orthodox acolyte near the church. He smiled when I asked the question through an interpreter-driver. Then, in rhythmic sing-song tones he voiced a strange, almost extinct dialect. His recitation was a prayer to the Virgin Mary.

So fragmented is Syrian society that no matter what course is taken by the party in power, it offends somebody. Enmities create desperate opposition. Coup follows coup. In Syria, it is not so much a case of power corrupting as it is of power disrupting.

Lebanon has its violence, too, mainly because of 160,000 or so Palestinian refugees sheltered within its borders. The Lebanese pay lip service to the anti-Israel sentiment of the Arab bloc. Many, especially the Christians, would just as soon avoid being drawn into that hassle, though the Moslem element is staunchly pro-Palestinian. So are many of the students at the American University of Beirut.

Periodically, there are riots as Arab refugees or pro-Palestinians demonstrate against Israel, seldom in a peaceful way, often with gunfire. Army and police struggle to maintain order, realizing that a mob may go berserk easily in this part of the world. Yet nothing seems to interfere with business, not even Sundays or holidays.

At the airport I protested one Sunday when a money changer deducted 10 percent for his commission when changing dollar traveler's checks into Lebanese pounds. "That's a terrible rate." I pointed to the much higher exchange rate listed in a Beirut city guide. "Look. Here's what it should be."

"I know," the clerk said, without changing expression. "Tomorrow I will give that rate. But this is Sunday and all the other banks are closed. So today I give this rate."

That was that. Charge what the traffic will bear, a typical reaction. In justice to them, they expect you to deal in the same way and will respect you for it. Where differences occur, bargaining starts, and bargaining is a way of life. This country imports six times as much as it exports, but lives handsomely by its capitalistic wits. It has no oil, coal, or any minerals worth exploiting. Its forests were stripped to make galleys for the pharaohs centuries ago. Its industrial base is small. There are fruit orchards along the lush coast at the foot of mountains rising to snow-covered peaks. But the total agricultural production contributes only 9.5 percent of the country's gross product.

Lebanon lives on commerce. It is the entrepôt center for the Middle East. It has seventy-three banks, thirty-seven of which are owned partly or entirely by foreigners. Second largest is the Russian-owned Moscow Narodny Bank. Russians like the easy Lebanese banking system, with its minimum regulation, its Swiss-type secret bank accounts, its free gold market, and its currency convertibility.

American, French, German, Swiss, British, and Latin American banks like this system, too, which is why you find all of these nationalities represented among the foreign-owned banks. Over a million Lebanese have migrated to foreign lands, including about 100,-000 who have settled in America. Ralph Nader, the dedicated crusader for the consumer, is the son of a Lebanese-born restaurateur. Emigrants dispatch more than $200 million a year to relatives in the homeland, often working for nearly a lifetime in America, Brazil, West Africa, or elsewhere. Most likely they will be running stores or operating commercial businesses. Then they may leave businesses to sons and retire to Lebanon.

Lebanese have a strong sense of humor. They can laugh at themselves, something not too common among Arabs. In Beirut, for instance, one hears the joke about America's first manned space shot to Mars. The United States government called for volunteers and guaranteed a certain sum of money to the family of anyone who would make the trip.

"How much do you think is fair?" a U.S. space agency official inquired of an American, the first volunteer. The answer was $100,-000.

A Frenchman was the next candidate. He wanted $200,000. "One hundred thousand dollars for my family and another one hundred thousand dollars for my mistress."

Then a Lebanese appeared. His figure was $300,000. He explained to the examiner: "A hundred thousand dollars for my family. A hundred thousand dollars is your cut, and the other hundred thousand dollars will go to the man we hire for the trip."

Everything in Lebanon seems to be tailored to foster the entrepreneur. Taxes are low. The Lebanese pound consistently is one of the strongest in the world. Lebanon itself is a sunny, pleasant land, in which the climate varies with the altitude, from tropical shoreland where bananas flourish, to cold Alpine heights where snow lies on ridges for six months of the year. Beirut, its capital, is a glittering city of over 700,000, with balconied high-rise apartments, plush nightclubs, and a seaside promenade along which big American-built automobiles speed at all hours. No other country in the Middle East offers its variety of scenery. No other city in the Middle East can match the sophistication and glamour of this capital of Beirut, where a substantial part of the area's business is handled.

Business booms because of oil money. While Lebanon may seem only a spectator in the oil playet, it, too, is an actor on the stage. Its banks transact business for oil-rich sheikhs. Its trucks haul merchandise to oil camps in Saudi Arabia, Kuwait, and the United Arab Emirates. Contracting firms service the entire area.

"In the contracting business, our volume rises along with the budget spending of Arab governments," explains Nadia El-Khourey. Her companies are building a pipeline and power station in Sharjah, a tank complex in Jordan, water tanks in Abu Dhabi, and various other facilities that help to get oil from the ground to tanker ports.

"There is no other area in the world that can produce surplus capital as fast as can the Middle East," says Saeb Jaroudi, curly-haired, athletic ex-minister for economy who resigned to head the Arab Development Bank in Kuwait. I encountered him in the bar of the Beirut Golf Club. He had just finished a round of eighteen holes and expressed intention to sharpen his "learner amateur" golf talents on the sandy links of Kuwait.

Jaroudi's job as head of the new Arab Development Bank calls for him to attract some of the oil money into area development projects. He is a go-getter with an M.A. from Columbia University who hopes that his new bank will become a sort of mini world bank of the Arab world. "There's no lack of money in this region. The problem is to put it to work," he said.

Imagine nations having so much money that they have difficulty putting funds to work! That is the situation in the Middle East today, and the money glut will worsen, if such a word may be used to describe a surplus of cash. Most nations and people do not suffer from this "affliction."

Lebanese do seem particularly adept at putting money to work, and scores of examples may be cited to prove the point. I first met Pierre Hélou in a penthouse office overlooking the sweep of Beirut's bay. Forty years old, he stood tall and erect behind the desk in his contemporary-styled office, curly hair slicked down and shining. He resembled the movie idea of the Madison Avenue advertising executive.

Cheerfully, Hélou told how he had started more than twenty business ventures in as many years. Half failed. Those that succeeded made him a millionaire several times over. I next met Hélou several years later when he had become Lebanon's Minister of State for Petroleum Affairs. The encounter was in the lobby of the Sahari Palace Hotel in Riyadh. He was leading a Lebanese delegation that was negotiating construction of a joint refinery in Lebanon with Saudi Arabian cash help.

This investment was in the fifty-to-seventy-million-dollar range and it was Saudi Arabia's first in downstream oil operations. This is significant, because Saudi Arabia will have far more money than it can spend at home in the years ahead. If surplus funds are not put to work, they may slosh around world money markets, upsetting weak currencies and that might include the United States dollar. The more money Saudi Arabia invests in downstream operations, the better it may be for America, Western Europe, and Japan.

I asked one friend in the Saudi Arabian government if he saw any significance in this particular development. He is a Texas-educated man in his early thirties, not given to much introspection, who looks for the obvious rather than the obtuse. He thought a moment over the question, then responded, "Well, for one thing you'll never go broke if you have a straight-shooting Lebanese for a partner."

Sure thing, pahdnah! One Lebanese proverb pays tribute to the entrepreneur. When a man is successful, neighbors are apt to say: "He is a man who can make a wine cellar from one grape."

A self-made millionaire is no rarity in Lebanon. One of them is Munir Abu Haidar, Chairman and owner of Trans-Mediterranean Airways. Twenty years ago Haidar was a $100-a-month clerk working for Aramco in Saudi Arabia. It is not unusual for a bright young Lebanese to take a job in Kuwait, in Saudi Arabia, or in one of the sheikhdoms of the gulf, working with an oil company, in a government bureaucracy, or with a trading company, and then, after he accumulates a stake, to go into business for himself.

Haidar had ideas too.

He noticed how oil companies in the area depended upon suppliers in distant points for that special lube oil, for diamond drilling bits, for food, and for the thousand and one things needed to maintain an oil camp. Sometimes shortages caught companies in an awkward position, leaving them vulnerable to slow ship schedules.

Why, wondered Haidar, could an air-cargo line not be established to service the oil companies? With $600 in capital, Haidar went into business for himself. As a start, he chartered a British York four-engined piston plane from Air Charter, in London, at $150 a flight. "I cleared $150 on the very first flight," says he, "and I was in business."

I met him in an office adorned with a lion-skin rug, and a sable antelope head on a wall. Haidar, short, athletically built, looks like Jack Armstrong the All-American boy several years later. He was tanned from a recent skiing trip. Black-rimmed spectacles rode on his nose. Lebanese businessmen favor Western business suits, often tailored in London or Rome. They may shop at Harrods in London or at Aux Trois Quartiers on the Boulevard de la Madeleine in Paris more often than at Papa Georges Bazaar in Beirut. And they are apt to be active even when hearts might call for a more restrained way of life.

Haidar likes to ride jumping horses, hunt big game, and ski on Lebanon's slopes in the mountains. He also likes to talk about his business and how he got his start.

The first month he had his air-cargo service, he operated twenty flights, averaging the same $150 profit a flight. This gave him a $3,000 profit in that initial month, enough to convince him that he had found something good. Before the year was out he bought his first plane, a York, for $15,000. Today, his operation is one of the largest all-cargo airlines outside the United States.

Equally impressive is Sheikh Najib Salim Alamuddin, President and Chairman of Middle East Airlines. At sixty-five he has a long record behind him, first as a mathematics and engineering professor at the American University of Beirut, then in various government posts, and finally as the man who forged MEA into a highly profitable airline without any government assistance.

"We celebrated when parliament did not go along with the idea of bringing state participation into Middle East Airlines," he said. "We prefer a free-enterprise airline."

This is a typical Lebanese reaction. Lebanon provides an interesting case study for those who believe that government intervention can settle any problem. It is in the heart of an area where governments do believe that the state should be all-powerful, operating under the guise of socialism. It has less in the way of resources than most states. Yet it thrives.

Sometimes the driving ambition of the Lebanese may be disadvantageous to an employer. Michel Doumit, an industrialist with interests in a cement company, the Bristol Hotel, and various other enterprises, says: "When you have a man working for you, he probably also is working for himself in some way."

This, of course, may be proof that free enterprise has served better than any other economic system yet devised by mankind. No other system so stimulates a person to work those extra hours, to devise methods of improving what is being done and to venture capital in speculative enterprises which may benefit everybody, not just the speculator.

George Azzi, a chauffeur working for a local businessman in Beirut, certainly understands the free-enterprise system, and benefits from it. He lives at Jiah, eighteen miles from the city on several acres of good farming land. Here he grows lettuce, tomatoes, and various other vegetables, using labor in the family to tend the crops.

On numerous days he uses his Chevrolet sedan of recent vintage to transport products to Beirut markets, where they are sold to wholesalers. Then he drives to a local garage to pick up his employer's automobile. His own Chevrolet is rented to a local taxicab company on a daily basis. So just before taking his employer's car, he turns his own over to the taxi firm.

As he works through the day, chauffeuring his employer about town, he is drawing his return from his own rented automobile. At the end of the day, he leaves his employer's car at the garage, picks up his own Chevrolet, and drives home. If he is not too late he may putter in his garden.

This isn't the work habit you encounter in the socialist states nearby except where the entrepreneur has been left alone to find his own economic salvation. Moreover, along with these work habits, there is more personal freedom than is found anywhere else in the Middle East, with the exception of Israel. This freedom includes a press that digs hard for stories and that freely reports every fact and rumor concerning developments in nearby nations.

"There are thirty-two daily newspapers in Lebanon, most of them published here in Beirut," said Samyr Souki, the chief of Business Services Research, a public-relations firm that has become the voice of some of the major business firms in Beirut. Because of the free press, oil companies like to locate listening posts and regional offices in Beirut. Here you can follow Baath Party developments in Syria and in Iraq; what the Russians are doing in North Rumaila Field near Basra, Iraq; the real meaning of shifts in government taking place in Egypt; the pace of oil shipments through Tapline to Sidon;

and where the oil is going when it leaves the terminal of IPC's nationalized line at Banias, Syria, or Tripoli, Lebanon, and how much oil Abu Dhabi might be producing at any specific time.

Beirut happens to be the air crossroads of the Middle East. Moreover, Lebanon's tourist business is as important to this nation as oil is to some of the Gulf sheikhdoms. Arab tourists from Saudi Arabia, the Gulf, Iraq, Syria, and Jordan make up the bulk of the arrivals and most of these travelers are business and oil men or government leaders who know what is happening within their particular country or orbit. Thus, at any one time a person in Beirut may encounter somebody from each of the other countries of the area.

Lebanon possesses something else that gives it a unique position in the Middle East—American University of Beirut. For over a hundred years, this American-supported school has been contributing teachers, doctors, engineers, scientists, and other educated people to the area. This university school started in 1866 as Syrian Protestant College, but church connections have long since been severed. In 1920, it adopted its present name.

Since its conception, American University has granted more than 17,000 degrees. Its current student body of 4,380 includes enrollees of both sexes from sixty-nine countries. The majority, however, come from the Middle East. Moreover, once educations are completed, 80 percent of the university's graduates remain in the Middle East to contribute talents toward developing the area.

Among AUB graduates have been three presidents of Middle East nations, ten prime ministers, more than thirty cabinet members, and more than thirty-five ambassadors. When the United Nations treaty was drawn in 1945 in San Francisco, nineteen graduates signed the document for their countries. This was the largest contingent from any university.

"We have been successful because we have made ourselves relevant to the area," Dr. Samuel B. Kirkwood, a Harvard medical-school graduate who is AUB's president, said. We sat in the spacious nineteenth-century living room of his residence on the campus with a half-dozen professors from representative departments. Dr. Kirkwood had insisted on bringing them together to explain just what relevancy meant. Dr. Kirkwood is a lean man of average build but he seems taller when he sprawls full length in a living room chair, holding a sheaf of papers which detail some of the university's accomplishments.

Classes are in English, in American-student-participation style. Of the 653 faculty members early in 1973, 118 were from the United States. Undoubtedly, the school offers one of the best possible forums for presenting the American way of life to Middle Easterners who may be the opinion makers and leaders of tomorrow. Yet it is not a propaganda school. Education is its prime function and it fills this role with regard to the needs of the area, rather than by waving the Stars and Stripes. The education has an applied flavor geared to those needs.

The university's school of agriculture has developed a new type of wheat, Najah wheat, which is adaptable to dry-land farming in the area. The cereal has doubled yields in some cases. Nutritionists have developed a high-protein food called laubina that is easily and cheaply manufactured from locally grown chick-peas and grains. Its engineering department has developed a concrete with imbedded glass fibers. Used as a building material, the concrete is a cheap replacement for steel, an expensive product in the Middle East.

"The university has a special role, a duty to train nationals in the efficient management of their oil resources," said Zuhayr Mikdashi, a professor of business administration who also happens to be one of the Middle East's top oil experts.

Mikdashi has the typical Lebanese respect for free-enterprise economics, and a way of presenting his facts that is convincing even when he keeps his voice in low key. Too often, of course, a raised voice is considered part of the argument, as if noise and truth are related, with the one necessitating an intensification of the other. Mikdashi makes it clear why Arab leaders have taken their positions concerning oil and why a university such as AUB should be helping to train the new managers of that oil.

He said, "The whole question of participation by governments in the ownership of oil companies was first voiced by Saudi Arabia's Oil Minister Yamani at a business seminar held on the AUB campus in 1968."

Seminars are a popular teaching method here for reaching beyond the undergraduate to business, industrial, medical, and governmental leaders who already are making their marks on jobs. A medical symposium, for instance, annually brings about six hundred medical authorities to Beirut from all parts of the Middle East for a four-day session. A dozen or so seminars in other topics are conducted every summer.

Dr. Craig S. Lichtenwainer, head of the university's medical school, which is probably the best in the Middle East, said, "Ours is a teaching hospital. We train doctors, nurses, technicians, public-health officials, and medical scientists for the whole area." He is a native of Seattle who went to Ethiopia on an aid project. There he found himself holding license No. 78 to practice in a country of fifteen million people. The thought of only one doctor for every 200,000 persons left such a deep impression on him that he has spent his life in the medical-training field.

Today the bulk of the students at AUB are Moslem, pro-Palestinian, and anti-Zionist. You seldom hear any anti-Jewish sentiment, and several students, when queried, emphasized that Beirut has a sizable Jewish community that lives unmolested in the heart of the city.

"We are not anti-Jewish. We are anti-Zionist. We favor a multiracial state in Palestine," insisted Mohammed Dajani-Dawudi, President of AUB's Student Council. Sitting in an office of the Student Union, he reported that he was a Palestinian himself, with parents still living in Jerusalem.

Zionism preaches that Israel should be a Jewish state, offering a home not only to its present citizens but to Jews from anywhere in the world who wish to leave the diaspora for aliyah in Israel. Palestinians charge that this immigration prompts Israel to take over more and more Arab land, reducing any inclination Israelis might have to return occupied territories.

Anti-Israeli bitterness probably reaches its height in Syria, an unhappy land that carries its *karamah* on its sleeve. An Arab's *karamah* is his dignity and his image, the appearance he presents to the world.

"It is extremely important for the Arab that he appears respected and of high status," says Elie A. Salem, political science professor at AUB. "Nothing is more devastating to the Arab than loss of face or humiliation in public. This accounts for the shame that Arabs felt as a result of the military defeat in the 1967 war with Israel." It played a role, too, in starting the 1973 Yom Kippur War.

Once an Arab's *karamah* has been injured, he is apt to react violently, often heedlessly, concerned only with striking back at his enemy and not about how his own injuries might be compounded. Syria sometimes seems to have a national *karamah* that has been hurt so many times that it seems ready to explode even without provocation.

When the Arabs first swept into Syria in A.D. 634 fired by their jihad, or holy war, they called the country "Bilad al-Sham," or "the country on the left." The name still fits. Syria is one of the most radical, leftist nations in the Middle East. It is a condition reached only after much instability and turmoil.

After French-"protected" Syria obtained independence in 1946, it considered merging with Iraq, gave up the idea, then fell victim to coup after coup. Instability is bound to be the norm when political opposition is unwilling to accept democratic verdicts. Trouble is compounded when the winning side is likely to have "fixed" that verdict, as was the case all too often in Syria. Usually the fix would be justified as an act benefiting the people. It does seem standard nearly everywhere for a winning party to aver that it has the people on its side. Syria has never failed to have a surplus of politicians and military leaders who claimed that they were on the side of the angels; angels just never appeared when their help was needed.

So it floundered through revolution and counterrevolution. An abortive attempt to merge with Egypt into the United Arab Republic lasted from February 1958 to September 28, 1961. A revolution on March 8, 1963, paved the way for the left-wing Baath Party to come to power two months later. This introduced a decade of intraparty bickering between moderate socialist and far-left wings of the Baath Party, with groups alternating in power through bloodletting coups. Nationalization of most segments of the economy only led to an exodus of vitally needed entrepreneurs, technicians, and businessmen.

In early 1973, General Hafez Assad, the Baath Party leader, sought to introduce stability through a new constitution. He has attempted to restore confidence through the use of more moderation than his predecessors. But this is an uphill struggle in a country as fragmented as Syria, with so many obligations to the Soviet Union which has been masquerading as one of Syria's few outside friends.

A crushing military defeat by Israel in the Yom Kippur War left the economy flat. Soviets did help launch a mini oil industry in Syria. But there is little oil in the country, and even that is of low quality. Production has been running at only about 130,000 barrels a day.

Syria's influence on oil has been more negative than positive. The two pipelines—one from Iraq and one from Saudi Arabia—that traverse its soil carry about 1.6 million barrels of oil daily when operating at capacity. That is equivalent to nearly 10 percent of America's daily requirements, enough to make quite a dent in world markets. Syria has stopped this flow several times when seeking more toll revenues from oil companies and when protesting against Western support for Israel.

Perhaps nothing illustrates the interlocking relationships of politics and oil in the Middle East better than does the rupture of Tapline's pipeline in Syria, on May 3, 1970. This break, in what appeared to be a deliberate "accident" on the part of Syrians, did more than anything else to emphasize that the oil companies' power in the Middle East had been broken, along with that section of 30/31-inch pipe of the 1,068-mile-long line. The details and some of its ramifications will be discussed later.

It should be pointed out here, however, that the pipeline break in Syria also served to emphasize how quickly the oil situation had changed in a few short years. The June 1967 Arab-Israeli war had closed the Suez Canal, a transport artery far more important than Tapline. Yet oil companies had been able to surmount this difficulty fairly easily by turning to giant oceangoing tankers for shipment of their oil. Egypt, the Middle East's most populous country, however, did suffer through loss of the canal. It found itself stripped of one of the few resources that gave it a foothold in international oil. It became even more of a frustrated nonentity in world oil economics.

XII

The Egyptians

The Central Bar is on Shari el Thawra, the main street in Ismailia on the Suez Canal. George Patyannis, the proprietor, stood erectly behind the bar, his trim frame showing that he didn't drink too much of his own Stella brand beer. He held his gray head high as if he might have been daring the Israelis to move him.

The bar occupied the lower floor of a two-story building that had somehow escaped the shell fire that marks so many other buildings here, where Egyptians face the Israelis. Nobody else was in the place: it did not look as if anyone would arrive in the foreseeable future.

In the old days sixty ships a day passed through the waterway. Off-duty crewmen and passengers would disembark at Port Said, Suez, or Ismailia to sightsee or to sport exotically with persons of easy virtue, male or female. Then, business boomed.

But why was Patyannis still here when no ships had passed since the June 1967 Arab-Israeli war closed the canal?

"This is my home," he said, simply. A sad smile played about his lips. A pitifully small row of bottles sat on a shelf behind him. One wall held a color photo of Lake Louise, Alberta; another had a color photo of the Parthenon in Athens.

He had some Greek ancestry. But he was Egyptian. That was evident in his ability to suffer without complaint, to accept life as allotted to him, and in his stubborn intention to remain rooted to the land of his birth. Some peoples of the world wear their national characteristics like suits of clothes for everyone to see. The national characteristic of the Egyptian is more abstract; it is the capacity to endure. He has done it for sixty centuries. If he does not breed himself to death with his 2.5 percent annual population increase, he may exist for another sixty centuries in his Nile River home.

Today there are thirty-five million Egyptians living in a dual society that is always strained by the pressure of population on limited resources, and now by war's aftermath. One society is that of the rural areas, where 60 percent of the population lives, tilling an average 0.7 of one acre per head. The other is the urban society where most of the literacy, wealth, and power are found.

Unfortunately, the oil wealth of the Middle East has missed Egypt. This country produced 220,000 barrels of oil a day in 1972, and consumed about 170,000. This left a driblet of net oil for export and gave it the credentials that finally allowed it to join the Organization of Petroleum Exporting Countries in January 1973. OPEC is the group that really controls oil in this part of the world. For all of its population and position as the leading nation in the Arab world, Egypt until now has had only a peripheral influence on oil policies in the Middle East.

In part this was due to geography and to the discriminating manner in which nature bestowed petroleum upon the countries of this area. It also was due to the devious politics of the Middle East. Egypt's President Nasser tried hard to forge oil nations into a bloc that would follow his leadership, using persuasion, then subversion, in Arab right-wing states. In the 1950s, the Arab League, an international organization in which Egypt always has had considerable influence, had a petroleum section, and it seemed as if this might become the vehicle to tie Arab oil producers together.

But when OPEC was formed in 1960, it became that vehicle. OPEC came into being in Baghdad, Iraq, at a time when Iraq and Egypt were vying for leadership of the Arab world. Iraq and other founder OPEC members cleverly worded the constitution so that only oil producers with net exports could become members. Egypt found itself outside what was to become the most powerful oil cartel ever created.

The Arab League, through its Arab petroleum congresses, played a minor political role in oil ever since. Sometimes its forums sounded like an empty drum, making much noise but not doing anything that mattered. OPEC had the power to squeeze more profit from companies for the producing countries. Middle East oil policy, therefore, was made in Baghdad, Riyadh, Kuwait, and Tehran, not in Cairo.

Even when Arab oil producers created their own Organization of Arab Petroleum Exporting Countries (OAPEC) in 1968, Egypt was an outsider. This organization seeks to stimulate coordination among Arab oil states through the establishment of a jointly owned tanker company, through creation of national oil companies that trade information and data about operations, and through other devices. In December 1971, OAPEC at a meeting in Abu Dhabi had to alter its constitution to permit Egypt and Syria to become members.

Prior to June, 1967, the Suez Canal had been Egypt's strongest instrument for affecting the oil economics of the area. In 1966—the last full year the canal was open—it handled 21,250 ships with a net tonnage of 274 million tons. About three-fourths of those ships were petroleum tankers, passing light to the Persian Gulf, and transporting crude oil on the upbound voyage. The 1967 war closed the canal, trapped fifteen ships on its waterways, and forced the international oil industry to realign its operations. The 1973 war left Israeli on both sides of the canal in places, with Egypt's economy in a perilous state.

Meanwhile, the average Egyptian resigned himself to harder times, whether he happened to be an urban or a rural dweller. Egypt's dual society shares the capacity to endure, though it may pool little else.

The sophisticated, Western-dressed Cairene may lunch at the Gezira Sporting Club while he scans his copy of *Al Ahram* to formulate his opinions. The *galabieh*-clad *fellah* may till his land with a buffalo-drawn plow, lift water from the Nile with a hand-operated *shadoof*, and grin shamefacedly through missing teeth when forced to admit an inability to read. But he may listen daily to a transistor radio tuned to Cairo Radio.

A vast gulf exists between literary and colloquial Arabic. Literary Arabic is the language of the Koran, sacrosanct and unchangeable. Colloquial Arabic has evolved from it like weeds in rich soil. The two are like Latin and Italian, or perhaps like the written English of Chaucer and the spoken English of Brooklyn. Thus, a poorly educated person may read simple Arabic yet be unable to comprehend a newspaper. So the spoken word is much more powerful in the Arab world than in the West. The radio reaches a far wider audience than does the printing press.

It is the curse of Egypt that outsiders sometimes judge the entire country by the *fellah*. They note his ignorance, and sense his discomfort when his *galabieh* is exchanged for an ill-fitting khaki uniform. Admittedly, the average Egyptian is not a good soldier, though he was a better one in 1973 than in 1967. He died bravely by the thousands, trying to regain land lost earlier to Israel. But he prefers to be left alone to till his soil, to fighting. He lives on little, seeks less, endures in silence, and breeds prolifically, for children mean more field hands. So outsiders look at him and say: "This is the Egyptian."

The *fellahin*, however, are only a part of Egypt. The cosmopolitan Cairene or Alexandrian holds power, and he may be shrewd and well informed. Typical is Harvard-educated Zaki Hashem, ex-Minister for Tourism, who has an easy charm that enables him to be at home in any society, including that of the girls of Radcliffe. Nostalgically he remembers when he and other Harvard undergraduates "chased Radcliffe girls."

As for women, only in Lebanon does the Arab woman have more freedom than she does in Egypt. Decades of contact with Europeans has given the Cairene and Alexandrian man a sophisticated outlook that enables him to view a woman as, if not quite his equal, at least as a human being with aspirations of her own.

"I think that this is due to the awakening of the Egyptian's dignity and of his growing desire to participate in things. Now women want to be treated as human beings too," said Leila Takla, an attractive professor of management at Cairo University. A lawyer by training, she is chic in a trouser suit, gold necklace around her throat, and bangles on her wrists.

There is even a women's-rights movement underway in Egypt, claims Amina el Said, the editor of the women's weekly magazine *She*, which has a circulation of 200,000. Its articles are geared for the modern female who wants to participate rather than function as a drone.

Women are being accepted in the professions and in government.

"Our government would come to a halt without women. Twenty-five percent of the 560,000 people employed by our government now are women," she said.

The Egyptian man increasingly accepts women as coworkers and his wife as an equal partner, she added. "Maybe at home he still keeps his superiority and is the boss, and we please men by giving in to them, or at least pretending to."

Among the *fellah*, the wind of change stemming from women's search for more rights may be less discernible, although sometimes he may be henpecked by a strong wife.

Such a one was Nagi Ebeid, a 61-year-old *fellah* who lived in Mit-Rahina, a Nile village twenty miles south of Cairo. Through an interpreter, he cheerfully admitted that he usually goes to bed shortly after sundown, to save kerosene in his lamp. Even though he grows some sugar cane, he cannot buy as much refined sugar as his six children, ages twelve to twenty-two, would like. But there are cabbages and turnips in his one-acre field. A cow provides milk. Forty date palms yield about 250 pounds of fruit per tree, annually.

He is a grizzled, small man, with a wife who outweighs and outshouts him and a two-story mudbrick house that is flush on the main street of the village. This street is a rutted, dirt lane on which only camels, donkeys, and pedestrians navigate.

A water-buffalo cow and calf were tied to Sayyid Ebeid's front doorpost, while a white donkey ambled along the street at our approach. The donkey followed the master into the house when the wooden door swung open on its leather hinges. Ebeid drove the animal outside, using quaint oaths and a sugar-cane stalk as stimulators.

Proudly he ushered us inside. Like most Arab houses, it was built patio style, with rooms facing an inner unroofed compound. An open fire glowed on the dirt of the compound. Beside it Ebeid's youngest daughter, a cross-eyed girl of about thirteen, laundered some cotton clothes in a *ticht*, a yard-wide pan that may be used for making bread, for washing dishes, or as a serving tray.

Sakina, Ebeid's wife, was a buxom, bossy woman in her fifties who was none too happy to have unannounced visitors calling. Though her name meant "peaceable one," she was anything but peaceable for the first few moments. Then her innate hospitality asserted itself.

She dropped the black shawl she had hastily lifted to cover her face at our entry, and her ultrapale features formed into repentant smile. She mumbled apologies about the state of her house. Nevertheless, everything was as neat as one might expect in a dirt-floored house where seven chickens and a duck were bedding in one corner and where the family donkey sometimes makes his home in the living room.

"My ancestors have been here since the time of the pharaohs," Ebeid said. He knew all about the pharaohs, whom he identified as the kings before Mohammed in that time of ignorance before Allah spoke to the world through the Prophet.

Perhaps he could trace his ancestry back to the time of the pharaohs. The *fellahin* of Egypt have been least affected by the invasions that have swept across this land ever since the time of the Ptolemies. The Egyptian peasant is more heavily built and more muscular than are the slender, bony Bedouins who still roam the deserts of this land. He is darker than the Arabs of Syria and Lebanon, perhaps a pigmentary step in the direction of the dark Nubians of the far south. He feels himself to be an Arab, and he is proud of the fact that for centuries Cairo and its Al Azhar University nurtured Arab culture after the Mongol invasions launched the long period of stagnation in the Arab world. Yet he remains an Egyptian, conscious of his own national culture, with a sense of national identity that is stronger than is found anywhere else in the Middle East, except perhaps in the state of Israel.

This is one reason why he is not willing to forget easily that his land is occupied by an alien invader. The Sinai peninsula is part of Egypt to him, even if it is sparsely populated. The oil of Abu Rodeis on the Suez Gulf is its only real asset and that now is in Israel's hands, along with the rest of the peninsula. This occupation is an insult to the Arab karamah. His political and military impotence fills him with a frustration that is almost visible in its intensity.

"We are being pushed toward war even though we would prefer a peaceful solution," Aziz Sidky, a member of Egypt's cabinet since 1956, stated one time when I encountered him in his office. At that time he was Minister for Industry and Petroleum, he had talked oil for an hour in a rational, modulated voice, providing a few asides about his education at the Universities of Oregon and Harvard.

Then Israel entered the conversation, and instantly his manner changed, his voice hardened, and he rose to his feet to stride a few steps across the carpet as if marching off to war with Israel. His frustration burst forth with a few bitter remarks about the way America supports Israel even though the latter occupies Egyptian territory.

"They say we won't sit down at the peace table and talk with them," he said. "But we have heard Golda Meir and other Israeli leaders say that they want to keep the Gaza Strip, Sharm el-Sheikh, all of Jerusalem, and the Golan Heights. So what is there to talk about?"

He is tall, with a shock of wavy hair and the authoritative manner of one who understands his job and is not afraid to make his views about it known. When I next saw him he was prime minister. Then in March, 1973, President Anwar Sadat, Nasser's successor, shook up his government. He took the prime minister's portfolio himself but retained Sidky as a presidential adviser.

President Sadat made this move in the spring of 1973 because he himself seemed to be a victim of the frustration felt by people in Egypt. The no-war-no-peace policy seemed to go nowhere, and Sadat decided that only war could change that, launched preparations for it in earnest. Until then, he had always seemed to be overshadowed by his predecessor, Nasser, a lightweight beside him.

Nasser had appealed to the sense of dignity that every Arab or Egyptian prizes. Even in his defeats, Nasser somehow managed to appear creditable, a victim of circumstances rather than his own errors, a man alone against the villains of Israel and America—or so he seemed to the average Arab in the Middle East.

Grandiose but futile political gestures only confirmed Nasser's greatness to Arabs. There was the abortive merger of Egypt and Syria into the United Arab Republic, the creation of an empty United Political Command for Arab nations, the rapprochment with the Soviet Union, which promised far more than it gave, and Arab summit meetings galore. Even the defeat in the Six Days War of June 1967 was not enough to unseat Nasser. His death on September 28, 1970, left a gap that still has not been filled, in Arab eyes.

Through most of his career, Nasser had his own personal public-relations spokesman in Mohammed Hassanein Heikal, the brilliant editor of *Al Ahram,* the Middle East's largest daily newspaper. No other man in Egypt has his knowledge of the complicated power web that exists in his country. No other man understands the nuances of that power quite as well, how an alliance in one area might affect happenings in another, how people might respond to a certain action on high.

At fifty, Heikal is a short, dark eyed, dynamo, with close-clipped hair, a square jaw which reflects his intense drive, and an obvious liking for the devious ways of Middle Eastern politics. Dressed in a finely tailored silk suit, Heikal not only is the Middle East's most important editor, he also looks it.

Al Ahram's headquarters in a twelve-story glass-and-concrete building in Cairo would be a revelation to anyone nurtured on the idea that the ignorant *fellah* is all that Egypt has to offer. Two hundred art works decorate its quiet halls and ornate offices. The newsroom is perhaps the least noisy in the world. There are no typewriters. Reporters write stories in longhand and give them to typists in another department who put them into Arabic script. Telephones in the newsroom do not ring, they blink. An electronic board on a wall provides instant notice of the progress of pages and sections of the newspaper into the finished product.

An editorial conference brings department heads to a highly polished round table of tropical wood in a soundproof room decorated with maps of the world. The discussion looks like a board meeting of a billion-dollar corporation. Editors in shirt sleeves or dark suit-coats ressemble successful corporation directors.

The entire building is a spotless, hygienic edifice that certainly seems to be what its editors claim it is—one of the most modern newspapers offices in the world if it is not *the* most modern. Stories from the typing department are fed into an IBM 360 Model 30 computer, which is programmed in Arabic, and handles 6,000 lines an hour.

Tape from the computer goes through electronic linotype machines at a rate of 1,000 lines an hour per machine. Presses roll at 360,000 copies an hour. With its electronic devices and gadgets, the plant has those presses rolling a half-hour after the editorial department deadline is reached.

Circulation is 600,000 with 200,000 of it going to nearby Arab countries. Moreover, with ad rates at $6.25 a column inch (in 1972), the paper, which belongs to the Arab Socialist Union, is a profitable enterprise indeed. Employees talk of the paper with pride. Satisfying that pride is part of the Arab *karamah*, of a man's search for dignity, of his desire to be treated as a human being. Man does not live by bread alone; he also needs the meat and potatoes of respect.

The late President Nasser understood this. He did redistribute some of the country's meager wealth to help the landless and the poor; but his real contribution was in bolstering the dignity of the individual Arab in Egypt and elsewhere. The camel raiser of Saudi Arabia, the trucker in Iraq, the oil rigger in Kuwait, and others would tune to Cairo Radio and could empathize with Nasser as he ranted against Western colonialism or Israel. It was easy to ridicule the gullibility of the listeners. Yet Nasser's emotions were tuned to the Arab character.

President Anwar Sadat has none of that charisma. His philosophies are on a solid Islamic base tinged with pragmatism. The closest he has come to playing the international leader has been in the move to federate Egypt, Libya, and Syria. There was no rabbit in this particular hat for Sadat. Egypt could use the wealth of oil-rich Libya, with its two-to-three-million-barrels-a-day oil output, depending upon what rate Mu'ammar el-Qadhafi, Libya's leader, establishes for it. But El-Qadhafi seemed more the leader in 1973 than Sadat. El-Qadhafi makes no secret of the fact that though he is willing to accept the number two position behind Sadat in any federation, he sees himself as eventually becoming number one.

"Egypt is a country without a leader and I am a leader without a country," he declared once when discussing his role in the politics of the area.

Egypt does sometimes seem to be a frustrated country without a leader, and it is not only Sadat's fault that this may be so. The inability to find any solution to Israel's occupation of the Sinai produces a festering sore in the consciousness of most Egyptians. That is a sore which grew more painful after failure of the Arabs to attain objectives in the Yom Kippur War. Sadat, like Nasser, was blaming America for his failure to achieve victory.

Sometimes the bitterness resulting from the political impasse is reflected in an anti-American tirade in editorial columns of newspapers or in a discussion across a desk. Usually this is directed against the American government rather than at any visitor to this basically friendly country. Egypt's oil comes mainly from two American oil companies that have partnership arrangements with the country's state-owned Egyptian General Petroleum Corporation.

The principal producer is Gulf Petroleum Company, a partnership between EGPC and a subsidiary of Standard Oil Company of Indiana. It operates the Morgan oil field in the Gulf of Suez, which produced an average of 165,000 barrels a day in 1972, operating offshore facilities, and it has found a new offshore field nearby which might be capable of producing 100,000 barrels of oil daily. Phillips Petroleum Company has a smaller operation in the Western Desert with EGPC.

The Egyptian oil ministry and EGPC officials scrupulously separate politics and oil. Nothing is allowed to interfere with production. Discontent with American government policy is never allowed to sour the businesslike relations established with American companies.

"Americans get things done. We like the way their companies operate," one Egyptian oil-ministry official once admitted to me. "We would like to have more of them working with us."

Egyptians keep hoping that new discoveries in the Western Desert may lift this nation into the ranks of moderate-sized petroleum exporters. But development of Egypt's oil has not been as rapid as had been anticipated a few years ago. This is the case, too, with an oil pipeline that is to be built from the Red Sea to the Mediterranean as a bypass to the dead Suez Canal.

On the Suez Canal, prior to the 1973 war, there was an air of ennui, of nothing much happening under the wary eyes of two opposing armies that face each other across the waterway. From the tower of the Suez Canal Authority in Ismailia, there was a sweeping view of Lake Timsa, a salty body of water joined to the canal. The dedication ceremony for the ten-story modern building had been held only a month before the June, 1967 war started, and many of the offices have never been occupied.

The elevators were not running. So it was necessary to walk up ten flights of stairs to a roof observation point. In Lake Timsa, the American tanker *Observer* swung on her anchor in a gentle wind.

It had been transporting a cargo of wheat to India when trapped as the last vessel in a downbound convoy.

Fourteen more ships were trapped in Great Bitter Lake, further south. The Suez Canal had been a one-way traffic lane: two "switches" were maintained so that upbound and downbound convoys could pass. One turnoff was the Ballah Bypass, between Ismailia and Kantara, a city on the northern end of the canal. The other was in Great Bitter Lake, on the southern end.

The road from Ismailia to Great Bitter Lake was a clutter of Egyptian Army traffic, tanks, trucks, troop carriers, and armored cars. A green line of palms to the left marked the Suez Canal and the armed border.

I had asked permission from the Farrell Line in New York City to go aboard their *African Glen,* the only American ship of the fourteen in Great Bitter Lake. They agreed, but the telex message about it relayed to me was a bit unsettling: "Farrell agreeable. Farrell says you may find it expensive to charter boat to get out, possibly a hundred dollars' charter. Also advise that shells have fallen in vicinity of ship, and you must proceed at your own risk, etc. Bon Voyage."

It wasn't the possibility of shelling that bothered me as we neared the lake. It was that little note about the "etc."

Actually, permission from the shipowners was not necessary. Like everything else in the battle zone, the ships on Great Bitter Lake were under Egyptain military command. Israel seized control of the lake in the 1973 war, but that was after my visit to the ships.

Cargoes included thousands of cases of beer, a considerable portion of one of Australia's apple crops, some New Zealand lamb, and many other products of Australasia. Standby crews have been taking care of the beer. Sailors now refer to the lake as being "thirty-five feet of water on top of five feet of empty beer bottles." Actually, Great Bitter Lake is misnamed. Its blue-green waters shimmer and glitter like a Florida inlet in bright sun. Crystal-clear depths invited swimming and spear fishing. Ships rode listlessly at anchor, several moored together to reduce need for standby crews, which have been acting as watch and maintenance men aboard.

Even Duffy, the mongrel mascot of the *African Glen,* drank beer.

He happily downed a quota of two bottles of Tuborg Danish brew daily, considerably fewer than the quota which pertained for the rest of the crew.

The fleet consisted of four British, two Polish, two German, one Bulgarian, one Czech, two Swedish, one French, and the one American freighter, with the *Observer* all by itself at Lake Timsa. Any hint of negotiations between Israel and Egypt raises hopes that perhaps ships may be freed as the first part of any agreement to reopen the Suez Canal. Hopes have been raised and dashed several times, but the 103-mile-long waterway remains sealed.

The international petroleum industry has long since concluded that the canal would be closed for a good while yet. Fortunately for the industry, companies had already launched a trend toward giant maritime tankers even before the June 1967 war. Closure of the canal accelerated that trend, and oil flowed smoothly to customers in Western Europe.

Higher prices, however, did add to Britain's balance-of-payments costs and probably accelerated the British pound-sterling devaluation of November 1967. Nevertheless, looking back, it is surprising how quickly the oil industry adjusted after closure of a waterway that up to 1967 had been considered absolutely vital to it.

The Very Large Crude Carriers, or VLCCs, as they are called, permitted that adjustment. These are ships of 200,000 deadweight tons or more. The world's largest, the *Globtik Tokyo*, is 483,000 dwt tons. By comparison, the *Queen Elizabeth II*, Britain's proud passenger ship, is only 65,863 dwt. But Shell has a 533,000-ton tanker scheduled for delivery in 1976.

The voyage from Abadan, Iran, on the Persian Gulf to London is 11,300 miles when ships sail around Africa's southern tip and only 6,500 miles via the open canal. In 1967, the canal was deep enough to accommodate ships of a top draft of thirty-eight feet. The giant ships now in operation have drafts of sixty to eighty feet when fully loaded, with some of the vessels being built likely to need a depth of one hundred feet to navigate. This means, of course, that the Suez Canal already is outmoded for fully loaded VLCC's.

United States Army Corps of Engineers data indicates how the economics of mass increases the efficiency of oil transport in VLCCs. The cost of transporting oil from the Middle East to the

United States eastern seaboard, for instance, is estimated at $5.35 a ton in a mammoth tanker of 326,000 deadweight tons. This compares with a cost of $9.40, using an older 70,000-ton ship that used to be termed a "super tanker." Other studies show that VLCCs of 200,000 tons or more may save around thirty cents a barrel compared with costs of shipping in smaller vessels.

Faced with such economics, Egypt hopes it may regain some of its lost prominence as a transshipper of oil in two ways. Plans have been drawn for a deepening and modernization of the canal when and if the Israelis withdraw from its banks. The cost might range from $500 million to $1 billion, depending upon how deep the canal might be, which might hinge upon Egypt's ability to obtain financing. Obviously, Middle Eastern nations will have enormous amounts of money in their treasuries in the years ahead. If Egypt could obtain some help from sister Arab nations, financing would be the least of the problems.

The canal hardly would be deepened enough to handle fully loaded VLCCs. But it might be able to handle downbound VLCCs when riding light on their way to the Gulf. This would save 4,800 miles of the voyage to the Gulf. Laden tankers then would sail around southern Africa.

With Egyptian and Israeli armies still confronting each other and with any canal-deepening project likely to take at least a few years or more, the picture of a thriving canal is now little more than a dream. Perhpas nearer reality is the SUMED Pipeline.

A consortium of European companies, including French, Italian, and British, had offered to build a 200-mile-long line from Ain Soukna, on the Red Sea, twenty-five miles south of Suez to the Nile River, under the river, then across the desert to Alexandria. Two parallel 42-inch-diameter lines would have an initial capacity of 60 million tons annually, rising to 120 million tons in later stages.

The European consortium procrastinated and an American group led by Kidder, Peabody, the New York City financial firm, won Egyptian support for the job just before the Yom Kippur War. That war, undoubteldy, has delayed things and the project must be studied in the light of new conditions. Originally viewed as a $250 million scheme, it is more likely to cost about $360 million in devalued dollars. Kidder, Peabody has offered to do the job with

American capital, with Bechtel as the contractor. Egyptians claim the project is economically feasible no matter who finally builds the pipeline.

This pipeline already has some competition from the 42-inch pipeline built by Israel from Eilat, on the Gulf of Aqaba, over 160 miles of Israeli territory to Ashkelon on the Mediterranean. The line has a capacity of over thirty-five million tons annually, with the likelihood of this being raised to sixty million tons a year. Most of the oil flowing over this line comes from Iran, because Arab nations do not want their oil to help Israel in any way.

As long as the confrontation with Israel continues, Egypt's future as a transshipper of oil is uncertain. In fact, Egypt's whole future is unpredictable in the aftermath of war. If occupied territories are not returned, another war is likely to come in another few years, say many Egyptians. This war might be followed by another, then another, if necessary, with those predictions being made not as threats, but with chilling fatalism.

In the Middle East it is easy to view this area as a political morass akin to Vietnam, in its own dehydrated way. Deserts replace jungle. The brown-gray Nile flows by instead of the Mekong. The characters are different. Yet the complexities that lie in the way of a peaceful settlement are akin to the intricacies that dogged America in Vietnam. I mentioned as much one time long ago when having late-afternoon drinks with an Egyptian businessman in the roof garden of the Semiramis Hotel, beside the Nile, though I didn't mention Vietnam itself since America hadn't yet been caught in that war.

Red-fezzed waiters in long, white robes were setting tables nearby for the anticipated dinner traffic. On the silvery Nile, a felucca with triangular sail worked upstream, a dark silhouette against the red twilight glare. I emphasized that I saw the Israeli-Arab situation as one of those diplomatic quagmires that seem to have no solution.

The businessman nodded. "Of course you know the old Arab tale of the frog and the scorpion?" he asked. I shook my head and he continued. "A scorpion once asked a frog for a ride across the Nile. 'But if I give you a ride you will sting me, and I would die,' protested the frog. 'Oh, but if you died I would drown,' answered the scorpion.

"The frog thought that over for a while, then said, 'Perhaps you are right. If you stung me you would be killing yourself.'

"So the frog let the scorpion climb onto his back and started across the wide river. In the middle of the stream the scorpion's natural instincts asserted themselves; he couldn't resist the temptation to sting the frog.

"As the frog was dying, he asked, 'Why did you sting me? Your action shows no sense. Now you will drown.'

" 'I know,' answered the scorpion. 'But remember, this is the Middle East.' "

I thought the story descriptive enough that I included it in a piece in the *Wall Street Journal* that appeared on August 15, 1958. Subsequently, that story was picked up by numerous other publications, often with the wording intact to describe the quagmire of the Middle East's politics.

Then, in 1972, I saw a story with a Saigon dateline, one that discussed the morass of Vietnamese politics. The story started with the tale of the frog and the scorpion. Only now, after the story had traveled halfway around the world, the river was the Mekong, not the Nile. And the final line was: "Remember, this is Vietnam."

The story fitted just as well in Vietnam as it did in the Middle East. The Palestinian, and Israel's unwillingness to admit that he even exists, is one reason why the Middle East and Vietnam have become synonymous as apparently insolvable problems. The Palestinian is not willing to disappear into the ranks of the Arab world as if Arab unity were something that really exists. His presence, with or without a gun in his hand, provides another dimension to the oil story that now is evolving.

XIII

The Palestinians and Jordanians

Mohammed Ali, an eighteen-year-old Arab guerrilla in a camou-
flaged khaki uniform, handed me a ball of plastic explosives, anxious
to show his knowledge of things military. This mountain encampment in an Arab state almost within sight of Israel lay among pines
high above a valley where heat waves shimmered in the bright sun.
But it was cool in this upland country where two dozen Palestine
Liberation commandos were undergoing a training course in sabo-
tage and guerrilla warfare.

It would have been pleasant if I had not been standing there with
a chunk of plastic explosives in my hand. It is like putty, soft and
malleable, with a chill feel that might have been partly psychological. Without a detonator, the material is harmless, or so Ali and
others in the group had assured me.

Several young commandos who seemed even younger than Ali's
eighteen years raced across an obstacle course, then dropped flat as
a hail of machine gun bullets passed over their heads. An instructor
sat behind a light machine gun on a rocky knob nearby, coolly laying
his barrage over the heads of the prone trainees.

"We try to be realistic in our training programs," explained the
one-time news editor of a Kuwait newspaper, who had been assigned
by the Palestine Liberation Organization to take me to this guerrilla
encampment. I had met him in Amman, Jordan, in the Inter-
Continental Hotel. He had extended a beefy paw, his round face
wreathed in smiles, and had cheerfully said, "Just call me Sam."

Just as cheerfully he had identified himself as a refugee from Jaffa, in what is now Israel, and a Cairo University graduate. He had quit a newspaper job in Kuwait to become affiliated with the PLO propaganda network.

This had been the start of several action-packed days as I was passed along through the guerrilla network to finally reach this encampment. Sam came along all the way, his eyes usually alight with amusement, as if he were vastly enjoying an adventure that did not seem to go with his plump frame.

Along this underground network I had met Colonel Abdulah Wajih, a graduate of Syria's military academy, who had been assigned to training PLO-affiliated *fedayeen*, or guerrilla, groups. He was a trim, athletic man who looked like Clark Gable in his prime, with a neatly clipped moustache and hair that was turning gray at the temples. He was born and raised in Nazareth, he said, but had fled his homeland in 1948 at twenty, just ahead of a Jewish military patrol. He joined the Syrian Army, and after winning a commission, had served with it until 1964, when he had been assigned to *fedayeen* tasks.

"We are fighting for the return of our land," he insisted during a discussion about the drives that compel a man to adopt terrorism as a means of expression. His eyes had clouded when I asked him if he had ever experienced any doubts when circumstances prompted him to dispatch some of his young boys on an errand of terror that might find them slaughtering innocent bystanders.

"There are no innocent bystanders," he said simply.

"But what about your own boys?" I inquired gently. "The attrition rate is high on missions into Israel. How does it feel to send *them* on a mission?"

His eyes stared at me for a long moment. He said nothing. Then his gaze swiveled toward the valley, where the far distance faded into the heat haze. "Over there is Palestine," he said, and I thought he had changed the subject. Never once in our conversations had he himself alluded to that land as "Israel." It was always "Palestine."

As his gaze swung back toward me, I noted the emotion on his strong features. He said, "We could live in peace with the Jews if our land were returned to us." His voice climbed higher as he emphatically added, "I am a Palestinian. My six children are Palestinian. We will be Palestinians as long as we live."

• Later in Israel I was to find that the word "Palestinian" was just as taboo in the Israeli vocabulary as "Israel" is among Palestinians. I had succeeded in obtaining an interview in Jerusalem with Golda Meir, Israel's prime minister. Warm and neighborly as a person, she is a tough and uncompromising leader when defending what she feels are Israel's rights. How, I asked, did she think the Palestinian refugee problem should be solved? It was a mischoice of words.

"Palestinian? There are no Palestinians," she exploded. Lighting another cigaret, about the fifth in as many minutes, she leaned across her mahogany desk in Israel's modern Parliament Building and forcefully launched a monologue. There were Arabs in this area, no Palestinians. When Jews were under fire, they had stood by their homes, defending them. "Arabs ran," she declared. "We did not chase them out. They left of their own free will. Again I say to you, there are no Palestinians."

Understandably, Israeli leaders cannot admit the existence of the Palestinian. Admission might imply that Palestinians, too, have some claim on the land of Israel. Yet refugees whether or not termed "Palestinians," pose one of the biggest barriers to settlement of the Israeli-Arab matter. As long as this confrontation exists, a dark cloud lies over the oil of the Middle East.

It is a negative influence, to be sure. The Palestinian Liberation Organization has spawned the militant Al Fatah. The latter has given birth to the Black September Movement, the guerrilla clan that staged a massacre at the Olympics in Munich, that murdered a Belgian and two American diplomats in the Sudan, and that plots other deadly strikes. At various times guerrillas have blown up oil pipelines. They threaten oil.

They have succeeded in enlisting most of the Arab governments in their cause. Saudi Arabia, Kuwait, and Libya contribute substantial sums from government revenues to Palestinian programs. Citizens throughout the Arab world donate on an individual basis. Slowly and steadily, Palestinians have worked to create a national spirit that now is readily apparent among Palestinians whether they happen to be working in an oil camp in Kuwait, at an automobile plant in West Germany, or in the bureaucracy of Saudi Arabia.

It is the tragedy of the Arab that he has always been able to unite much faster against than for something. The Palestinian, by firing his hatred of Zionism, has turned that heat inward to fuse his own nationalism. Now it is too late to claim that the Palestinian does not exist, just as it is too late for anyone to ignore Israel.

Today, there are approximately 2.7 million Palestinians, according to PLO sources. They include 750,000 in Jordan, or about half the total population; 620,000 in the West Bank, the area west of the Jordan River to the pre-1967-war borders of Israel, an area now occupied by Israel; 380,000 in the Gaza Strip, the sliver of land just south of Israel's pre-1967 borders; 450,000 in Israel, most with Israeli citizenship; 180,000 in Syria; 160,000 in Lebanon; over 100,-000 in Kuwait; and the remainder scattered in the Gulf states, in Egypt, and elsewhere.

That total includes perhaps 1.3 million Arab refugees who fled from their homes, either in 1948 when Israel was established, or in the June 1967 War. The figure is controversial. The United Nations Relief and Works Administration, which helps support refugees, listed 1,471,220 on its rolls in 1971. This may be somewhat inflated. Nevertheless, the 726,000 refugees of 1948, plus those of 1967, plus the natural increase, probably adds up to a figure close to 1.3 million.

The history of the partition of what had been British-mandated Palestine is clear enough, though one may find himself in a can of worms when seeking moral justification for some of the atrocities allegedly committed by both sides in this long confrontation.

This land had been Arab since about A.D. 635, with Crusader overlordship for a century and Turkish rule from the fifteenth century to World War I. Jews had been dispersed from their homeland in A.D. 70 by the Roman general, and later emperor, Titus. Nevertheless, there have always been some Jews in the area, who through the centuries were joined by a trickle of immigrants from other lands.

Then in the late nineteenth century Zionism arose as a force, with Zionists claiming that Jews have a historic right to their old homeland in the Land of Canaan. Theodor Herzl, long the leader of the Zionists, defined its aims at the Basel Congress of 1897: "Zionism strives to create for the Jewish people a home in Palestine secured by public law."

Unfortunately, the bulk of the population in Palestine at that time was Arab. The trickle of Jewish immigration, however, became a steady flow through most of the first half of the twentieth century. Early in 1948, the population of Palestine was estimated at 2,-065,000, including 1,415,000 Arabs, or 69 percent. This included the West Bank and Gaza, so it would be wrong to conclude that all of these Arabs were competing directly with Jews at that time for the land. Figures do show, though, that any multiracial state created at that time certainly would have been predominantly Arab.

Israel's Central Bureau of Statistics lists the new nation's population at 873,000 in November 1948, with 156,000 of them Arabs and Druses. The bulk of the Arab population had flown. Israel definitely was a Jewish state under its new constitution.

Again there is controversy concerning the cause of that refugee stream. Arabs claim intimidation and force led to mass abandonment by Arabs of their homes. Israelis contend that Arabs fled of their own free will, or in response to radio urgings of Arab leaders in lands that then were attacking the new Israel.

Nobody denies that the vacant Arab lands and homes were expropriated by Israel. Everybody admits that the Arab refugees never received a penny in compensation for their losses.

Thus was born the Palestinian, one more artificial creation in an area where all too often nations have been established by diplomatic map makers as political whims moved pencils across charts. The Palestinians were created as a people before they had a state they could call their own. And it is no good to say that, being an artificial creation, the Palestinians have less existence than has a people forged over a long period of time.

The Palestinians' sense of nationhood developed as a mass homesickness that was never allowed to die. Ironically, this nationhood was shaped in a few decades by some of the same forces that helped to keep the Jewish nation alive through centuries of persecution. Just as injustices drove Jews closer together in their ghettos, a sense of injustice bound Palestinians together like mesons binding the particles of an atom. The refugee camp in the Gaza Strip, in Jordan, in Syria, or in Lebanon became both the ghetto and the school for this new nationalism.

Arab nations nearby unwittingly contributed to the formation of the Palestinian nationalism, too. For political and economic reasons they found it difficult to absorb the refugees. Generally, nations were happy to leave refugees in their ghetto-like camps. But it is a fallacy to believe that any of the contiguous Arab countries easily might have resettled the Arab refugees.

Bayard Dodge, one of the real American pioneers in the Middle East, spent thirty-five years of his life, from 1913 to 1948, as teacher, administrator, and president of the American University of Beirut. After leaving AUB he lectured on Arab affairs at Columbia and Princeton universities, wrote several books, and served as a Middle East adviser to the United States Department of State.

Just before he died at eighty-four, Dodge was asked the question, Why can't other Arab countries absorb the refugees? It was asked in an interview published in the July–August 1972 issue of *Aramco World* magazine, the publication of Arabian American Oil Company, in Dhahran, Saudi Arabia. Dodge gave this reply:

> People don't know very much about the Arab countries. You see, Egypt is terribly overpopulated. She can't take very many refugees, in fact, she can't take any. And little Lebanon already has over 100,000 Arab refugees down there around Tyre and Sidon. They really don't have room for any more. Jordan, of course, has most of the refugees, but Jordan can barely support her own population. So the only country where some of them might do something is up in Syria. But Syria has its own problems. When the French left, they didn't leave a good civil service and poor Syria has been going through one *coup d'etat* after another. So, although Syria could perhaps be a fertile, rich country, they haven't been able to develop the country so as to be able to take in refugees.

Lebanon has denied citizenship to the Arab refugees it does have, and Lebanese politicians say that it is lucky for peace in the area that this is so. The reason is that this little country is divided equally between Christians and Arabs. An influx of new Moslem citizens would tip the power balance toward the Moslems. The latter would push the nation toward belligerency vis-à-vis Israel, even though this might be suicidal.

Syria also is fractionalized politically and is not anxious to add bitter Palestinians as citizens, people who would be more interested in righting their own personal wrongs than in helping build a stable Syria. Egypt has all it can do to support its own population. Kuwait has over 100,000 Palestinians already working in the country, and risks being overwhelmed by non-Kuwaitis should it open immigration doors any further. And so it goes.

Only Jordan has extended citizenship to any refugee who wants it. But Jordan is the twin of Palestine, and might even find itself as part of any new "Palestine" that might emerge from the tangled politics of the Middle East.

In the 1948 war, Jordan annexed that part of Palestine known as the West Bank. Jordan became a two-part kingdom. On the East Bank, Bedouin power preserved the throne. On the West Bank, Palestinians paid nominal allegiance to the king while bitterly watching Israel. The 1967 war changed that, when Israel occupied the West Bank, and another 200,000 refugees fled to Jordan.

"We had to offer shelter to our brothers," said King Hussein, Jordan's 37-year-old monarch.

The UNRRA handles the relief job, with the United States picking up most of the tab. Its contributions have totaled about a half-billion dollars since 1948.

I met the king in his functional Basman Palace, which sits on one of Amman's seven hills. He greeted me at the door of his office as two Circassian guards in long, blue Cossack coats looked on. The king was dressed in a well-tailored suit, and wearing Italian boots. He is a trim, 154-pounder, deep-chested and muscular, perhaps about five and a half feet tall. Sometimes he is called "the little King." Close up, with his military bearing, he gave no impression of smallness. He is handsome, with a pencil moustache, an easy smile, and a strong interest in anything mechanical or athletic.

He apologized profusely for the delay as he led me past a desk cluttered with models of airplanes to a gold satin-covered divan. The walls of the office were in pale-gold velvet and mahogany. A Persian carpet lay on the floor.

I am not used to hearing a king apologize. I began to feel like royalty myself, a lower-echelon prince perhaps. When the king offered me a Chesterfield, I was ready to take one, even though I do not smoke.

As I sat down I remembered what one of his aides had said earlier: "He can charm you into saying 'yes' when you mean 'no' or 'maybe.' "

Appropriately enough, he elected to talk about Arab propaganda and its weaknesses. "We have been our own worst enemies in this regard. The Arab case has never been presented adequately in the Western world. Arab propagandists have made many mistakes—many, many mistakes."

His own flair for public relations failed to charm Palestinians. Squalid years in camps bred a desperation among them that led to development of the *fedayeen*, the Arab guerrillas who adopted terror as their avenue for striking back at Israel and for calling the world's attention to their plight.

Thousands of refugees have been living in camps ever since 1948. One Jordanian government study estimates that the value of rations distributed to families averages about $20.40 per person annually, or about 5½ cents a day each. Total relief costs amount to about 12 cents a day per person.

Obviously, relief and welfare payments are not substantial enough to hold refugees to camps if they have any place to go. The old and the very young do not, but young adults scatter throughout the Middle East, taking jobs in Kuwait, Saudi Arabia, Oman, Lebanon, and even as far away as West Germany.

Palestinians have developed a passion for education, realizing that this may be the only means of rising above the misery of a slumlike camp. Thus, today you find many of them holding good jobs in the area. Nearly all of the professors at the University of Amman are Palestinian one-time refugees. King Hussein's bureaucracy is interlaced with them in the better jobs, the director of agriculture at Abu Dhabi's Buraimi Oasis is a Palestinian, and Aramco has dozens of them in key posts in Saudi Arabia.

Their earnings help support brothers, parents, uncles, cousins, and distant relatives in camps. Family ties are strong and the Arab who has something may be supporting a dozen relatives who have nothing. The job-holding Palestinian also probably is supporting the

Palestinian Liberation Organization and its guerrilla offshoots such as Al Fatah. He may work awhile, then surrender to frustration and join one of the violently radical groups himself.

And always in the camps, the anti-Israeli propaganda churns away, creating new guerrilla recruits. At nine or ten, young boys in camp may learn how to fire rifles they can barely lift. At fourteen or fifteen, they may join a guerrilla group.

As the years have passed some of the camps no longer are camps in the strict sense of the word. They are small towns. New Camp, near Amman, is not a new camp at all. It is twenty-four years old, looking just about like any other town in Jordan of 25,000 population.

Rows of mudbrick houses, each built compound style to enclose small gardens, stand beside dusty streets. A two-story concrete school is being flushed with a water hose by a janitor. Several trucks loaded with watermelons are in a market area, drivers selling the fruit from back ends of vehicles. There is a business district with an open-fronted barber shop, a tinsmith's shop and several grocery stores. Bolts of bright cotton goods lie on a mat in the street. Women in long, flowing dresses and head scarves bend down to feel the goods and to check patterns. Men in *keffiyehs* shuffle along streets.

"Certainly, this is a city," says Akman Shehadeh, the brawny manager of the camp, who might have been the mayor were his job elective. Proudly he leads the way into a small, three-room building that serves as the "city hall." Standing at a blackboard, he uses a piece of chalk to illustrate how he administers the camp, operating through six departments: administration, health, education, sanitation, supplementary feeding, and welfare. His departments have 250 employees in all. The largest number, 200 teachers, is employed in the six schools of the camp.

On a wall in one of the schools, an artist has painted a mural of a Palestinian mother with child. In English, lettered wording under the mural starts this way:

Before you go
Have a minute to spare
to hear a word on Palestine
and perhaps to help right a wrong.

The story continues with the tale of an Arab Palestine that was stolen by a race from overseas, with Arabs driven out as refugees. The story ends as follows:

Today there are a million of us;
some like me but many like my mother
wasting lives in exiled misery,
wanting to go home.

The tale is more restrained than most of those presented to school-children in Arab refugee camps. From their first day in school they are taught that Zionism is the real enemy, that it is a treacherous force that is seeking to take all Arab land for a Greater Israel stretching from the Nile to the Euphrates River and that the wrongs inflicted upon Arabs must be rectified with extreme violence, if necessary. And children are left in no doubt that violence probably will be necessary.

This was the approach taken by the leadership of the PLO too, when Ahmed Shukairy was its chairman. He is a graduate of the American University of Beirut who made a career for himself as a Jordanian lawyer before going to the United Nations to represent Jordan. Then, in 1963–67 he was chairman of PLO. He still held that position when I met him in Cairo, just before start of the June 1967 war.

I had been told that an anti-Israel public rally was to be held at Bulaq el Guedid Street and that Shukairy would make an appearance. If I grabbed his sleeve, I might be able to catch his attention and perhaps even obtain an interview.

The street had been blocked to traffic. Straw mats had been placed on the pavement. On these, thousands of people already waited in the evening heat, fanning themselves with magazines or with straw fans. An Egyptian National Guard unit served as ushers. I worked my way to the stage where a string of colored lights hung over the street. Huge pictures of President Nasser hung from apartment buildings on each side of the street, along with banners in Arabic characters. They read, "Death to Israel."

I showed my press credentials to youthful guards and was led onto the stage where perhaps two dozen dignitaries sat on folding chairs near a microphone waiting for the rally to start. A wildly gesticulating fellow leaped to his feet from the crowd, jumped onto the stage and yelled, *"Ahahdoon! Ahahdoon!"*

This was the "we will return" cry of the Palestinian. Immediately, the crowd picked up the chant. Like students following a cheer leader at a football game, they repeated the call again and again. A rhythmic clapping of the hands started. One could feel the war fever of the mob, sense the smell of blood in the air.

I worked my way to a seat, found myself sitting between Shukairy, the suave and distinguished-looking chief of the PLO on one side, and Mohsina Tawfik, a dark-haired, attractive young woman who is one of Egypt's most noted film stars. Shukairy showed his surprise when I started plying him with questions in American-accented English. Then he answered in excellent English, an amused smile about his lips.

"Certainly there will be war," he said. "We will drive the Israeli into the sea."

Then Miss Tawfik sprang to her feet as if moved by a sudden impulse. But I noted that she had a script in her hand when she grabbed the microphone and clutched it against her. The crowd emitted a roar. Even before it died down she started her tirade, shoulders shaking as she passionately read from the script.

"What is she saying?" I asked of the Cairo Radio announcer who had slipped into her seat.

"She says 'American imperialism is against all peoples of the world. Americans are the enemies of peace.' "

The crowd roared approval. Shukairy raised his voice and leaned foward to say, "You Americans had better start worrying about your oil investments in the Middle East. Your support for Israel may prove expensive in the long run."

I didn't say anything. Miss Tawfik was continuing her tirade.

My interpreter's apologies were profuse before he said, "She says, 'Long live the good people of North Vietnam. Death to American invaders.' "

"Oh," I said.

"Forgive me. I am only repeating what she said," my new friend said.

I nodded, then saw a boy of about eleven climb onto the stage near me. I lifted a hand, helped him up when he stumbled on some wiring. The boy seized the microphone as Miss Tawfik released it and began to yell in Arabic, "Kill Jews! Kill Jews! Kill Jews!"

The crowd stood up, took up the cry. The streets reverberated with the shouts. There was a scuffle in the crowd, then another. A National Guardman brought the butt of his rifle down hard on someone's head. Hysteria was sweeping through the mob.

"Kill Jews! Kill Jews!"

The cry seemed to mount even higher when Shukairy strode to the microphone. With all eyes focused on him, I took the opportunity to slip through a rent in the canvas awning back of the stage and jumped to the ground. I found myself in a dark alley of the city, with the stage blocking off most of the light. As I hurried away I heard the shouts changing again to *Ahahdoon, Ahahdoon.*

The next time I saw Shukairy was after the June 1967 war. He was in eclipse, and Yasser Arafat was emerging as the Al Fatah chieftain and the dominant figure in the organization, and there was a subtle difference in most of the propaganda. Zionism was the enemy now, not Jews per se. There was no more talk about "driving the Jews into the sea." Now the emphasis was on creating a multiracial society in a Greater Palestine that would envelop Israel. America's government still was the enemy but threats against oil were muted. The Arabs had tried the boycott route immediately following the June war but found that Gulf producers paid only lip service to it. America then did not need Arab oil directly. Britain, the other boycott target, managed with Iranian oil and with oil that it somehow obtained from some of the Arab countries, though it did have to devalue its pound sterling. Arab nations like Kuwait, which held money reserves in London, were also affected.

The collapse of the Arab armies in June 1967 gave new life to the guerrilla movements. Leaders like Arafat and George Habash cried that the war had to be put on a guerrilla basis. Every Arab would become part of the force battling Israel in every way, creating a new Vietnam in the Middle East.

The trouble was that the arid borders of Israel provided little cover for guerrilla fighters. Israel met every foray with massive retaliation. Frustrated guerrillas built up their strength in Jordan and began to feel that they possessed more power than did the army of King Hussein. Splinter groups proliferated, and sometimes each seemed

to be vying to be further to the left in politics than the other. Al
Fatah had competition from the Popular Front for the Liberation
of Palestine (PFLP), a Marxist group that received Russian and
Communist Chinese support, from Al Saiqa, a leftist group nurtured
by the Syrian Army and others.

Ghassan Kanafani, a one-time journalist who had gone to school
at American University of Beirut, represented the type of guerrilla
emerging. I met him in a small apartment building on a narrow
street in Beirut. Revolutionary posters papered walls of his cluttered
office. Che Guevara frowned from one of them, cigar held like an
upward-pointing pistol in his hand. In another poster, Ho Chi Minh
stared straight ahead, as if seeing the final victory of his North
Vietnamese forces.

A comely brunette dressed in army fatigues met my smile with a
frown. She gripped her Kalashnikov automatic rifle a little harder
and glanced away. Kanafani was small of build, neatly dressed in
sport shirt and slacks, a heavy moustache adding little age to his
youthful features. He looked like a college freshman who has found
for the first time that he can grow a moustache, then tries to build
a personality around it. There wasn't anything collegiate about his
manner, though.

"We are a political movement as much as a military force," he
said. "This is where our real power lies." A hand grenade lay atop
a sheaf of papers on his desk, as if serving as a paperweight. His
female guard had a few more grenades dangling from a belt around
her slender waist.

He made it clear that the PFLP, with which he was associated,
aimed at overthrowing reactionary Arab regimes as well as Israel.
From the way he talked it was evident that just about all the Arab
regimes were considered to be unprogressive according to the standards of Marx, Guevara, and Ho Chi Minh. These standards dominated the political philosophies of this guerrilla group. But the most
reactionary regime in the view of Kanafani was that of King Hussein
in Jordan.

Kanafani was to die from a bomb planted underneath his automobile in a Beirut street, a victim, PFLP sources said, of Israeli secret
agents. Before that time, guerrillas had reached the stage in both
Lebanon and Jordan where they considered themselves above gov-

ernments. In bloody clashes, Lebanese forces maintained control, preventing establishment of the Lebanese border as a front for guerrilla war with Israel. Lebanon feared that Israeli retaliations would cause far more damage to Lebanon than would guerrilla raids into Israel. Massive retaliatory raids by Israel did show this to be a logical analysis of the situation.

But in Jordan, guerrillas grew bold enough to challenge King Hussein himself. They established their own road blocks to control traffic. A tax system was established, with guerrillas doing their collecting while toting their Kalashnikov rifles. They showed their contempt for King Hussein's Bedouin-based army.

In September 1970, the PFLP scored what seemed to be a triumph when four commercial airliners were hijacked. This coincided with an attempt to assassinate the king. One of the hijacked liners, a Trans-World Airlines jumbo jet, landed on a remote airfield in Jordan and was destroyed. Shortly after, King Hussein moved against guerrillas with his full army strength. Bitter fighting raged in Amman and elsewhere. Several hundred died, and the power of the guerrillas in Jordan was broken.

These attacks in September provided the reason for the Black September Movement, which later was established for a dual purpose: to avenge those events of September 1970 insofar as King Hussein was concerned, and to utilize terror to keep the cause of the Palestinians before the world.

King Hussein, a courageous man who survived innumerable assassination attempts, still retained his throne in 1973, though nobody can predict in the volatile Middle East how long he might survive. He was only indirectly involved in the Yom Kippur War, luckily for him from an Israeli standpoint. But this displeased some Arabs.

Jordan's ruler, unfortunately, has no oil to provide the financial backing for his government. Tapline, the pipeline from Saudi Arabia to the Mediterranean, does traverse Jordanian territory. Its pipeline fees provide some oil revenues, but not much by comparison with what neighbors earn from oil.

Palestinian guerrillas have never been very successful in striking at Western petroleum companies through oil, though they have blown up pipelines at different times. As long as governments allow repairs, these can be effected in a matter of days. Thus, such activities represent economic pinpricks of little real consequence, though as the world oil pinch grows any disruptions at all could cause economic damage in America and Europe.

A pipeline rupture backed by an Arab government may be serious at any time. That Tapline rupture on May 3, 1970, in Syria indicates the domino effect which is possible today in international oil. One small break in a pipeline on bleak plains in southern Syria played an important role in adding three billion dollars to the bill oil companies had to pay Persian Gulf producing countries in 1971. And that was only part of the total cost of this whole chain of events.

Until that black May day, Tapline had been transporting about 500,000 barrels a day of Saudi Arabian crude oil to a terminal at Sidon, Lebanon, where the oil was transferred to tankers for distribution, mainly to Western Europe. This was an efficient oil transport system that saved the long haul around the southern tip of Africa. With the Suez Canal closed, the southern route remained the only way for Persian Gulf oil to reach European markets via tankers.

Then, in early May, a bulldozer cut the pipeline. It was an accident, Syrians said, but this pipeline, buried in a shallow trench, snakes across the open country like a giant hose with an outer crust of sand. It would have taken a stupid bulldozer driver, indeed, to scoop out a section of the pipeline. Moreover, Palestinian guerrillas had been operating in that region for months with the blessing of the Syrian military.

Suspicions hardened when Syrians refused to permit Tapline technicians into the country to repair the pipeline break. A rupture that might normally have cost a shutdown of only a few days became an international incident with worldwide implications.

Now its ramifications were to lead to Libya, then into the London tanker charter market, back to Libya and delicate negotiations with oil companies, and eventually to a profit squeeze on those companies in Tehran, Iran, in February 1971.

Libya was led by radical, erratic Mu'ammar el-Qadhafi, the army officer who was only twenty-seven in September 1969, when he led the coup that overthrew the aged King Idris. Oil production in 1969 averaged 3.2 million barrels a day, and Libya seemed destined to challenge Iran and Saudi Arabia for leadership of oil production in this part of the world. But el-Qadhafi, a shrewd intriguer when dealing with oil companies, had other ideas. Shortly after taking power, he opened negotiations with companies seeking a better deal than the 50–50 profit split between companies and the host country, which generally applied.

Negotiations dragged into the spring of 1970. There was an adequate supply of petroleum on world markets, and the oil companies were under no great pressure to settle. The companies operating in Libya included American Overseas Petroleum, Ltd., British Petroleum Company, Ltd., Société Nationale des Pétroles d'Aquitaine, Exxon, Mobiloil, Gelsenberg, Oasis Oil Company, Occidental, and Nelson Bunker Hunt.

The oil companies had started 1970 with a big upsurge in production. Then el-Qadhafi turned down the valves and ordered production cutbacks, ostensibly to conserve Libya's petroleum resources. Actually, the move seemed to be another way of putting pressure on the companies. As for el-Qadhafi, he preferred a high per-barrel price for a lower volume of oil than higher output at a lower price.

Still the companies did not feel the hurt. Oil was flowing from Iraq and Saudi Arabia through two pipelines at a 1.6 million-barrels-a-day rate to the eastern Mediterranean. Exxon, the biggest Libyan producer, and Mobiloil were sharing in the Tapline oil from Saudi Arabia as well as working separately in Libya.

Then came the Tapline rupture in Syria. Normally, companies might have hiked their Libyan production immediately to substitute for the 500,000 barrels a day of lost Saudi Arabian crude, which flowed via Tapline.

"This would have been the logical thing to do in ordinary circumstances," said Tom Chandler, President of Tapline and a long-time veteran of Middle East oil. Sitting in his Beirut office on Rue Hamra, he seemed to possess an internal crystal ball as he explained why this would not be happening, then predicted what this might mean to the Western world. The term "energy gap" had not been coined yet. But Chandler saw it coming.

Instead of permitting production rises, Libya's el-Qadhafi forced more cutbacks. Overnight the flow of oil to Western Europe changed from surplus to shortage. This is a grim reminder of what might happen in the future if politics again intrudes in oil when demand from America and Western Europe may be much higher than it was in 1970.

The oil corporations could have increased the production of subsidiaries in the Persian Gulf to make up the deficit in supply going to Europe. But with the Suez Canal closed, it meant that additional oil—500,000 barrels a day of lost Tapline oil plus twice that volume of Libyan oil cutbacks—had to be transported all the way around southern Africa to market.

In London, where ship charters are arranged, tanker charter rates soared as every major oil company fought to obtain tanker tonnage. Freight rates reached record levels. As pressure on the companies mounted, Libya increased its demands, seeking higher posted oil prices and an increase in the tax rate from 50 to 55 percent.

The posted price for oil is fictitious. In the old days it might have meant something, though even then, during times of sluggish demand, the companies might shave the posted price in special deals to attract customers. Thus at times a posted price might be $2.75 a barrel when oil really is selling at $2.30. Meanwhile, the posted price would remain as the "official price" that others might have to pay. Since the taxes of producing countries are based on the posted price, these nations had a vested interest in keeping the posted price high.

The system offered one more ingenious way for countries to squeeze oil companies. Once having reached a 55 percent rate, countries were not interested in dropping back to 50 percent. They became convinced that if oil companies could absorb the 55 percent rate in competitive times, the firms should be able to absorb the same rate in good times.

Libya was the leader in pushing for the higher rate. As the oil pinch mounted in Europe, it pressured companies separately to agree to the new taxes. Occidental Petroleum Corporation, which depends upon Libya for much of its oil, was the first to meet Libya's terms. Its chief executive officer, colorful Dr. Armand Hammer, is somewhat of a maverick, an individualistic entrepreneur who makes decisions like a tycoon of the days of industrial empire building in America. After he led the way, other companies found themselves forced to surrender to Libya one at a time.

The Libya victory stimulated fresh demands on companies else-where. Oil nations jealously watch every deal negotiated by oil com-panies with "brother" countries. Just as one labor union thinks it should obtain every benefit negotiated by a brother union, oil-pro-ducing nations feel that if oil companies give anything to a particular country, other producing nations should have the same benefits. Thus oil companies were forced to accept the same 55 percent tax in other countries.

The oil nations sensed that market conditions were swinging in their favor. In December 1970, OPEC decided in Caracas, Venezuela, that posted prices of Persian Gulf oil should be in-creased, raising tax takes, too. Iran, wanting more money for eco-nomic development, led a determined OPEC group in negotiating sessions with company representatives in Tehran through January and into February 1971.

Faced by a shutdown of petroleum production in all OPEC na-tions, oil companies surrendered. On February 14, 1971, they as-sented to what has become known as the Tehran Agreement. It provided for a 33-cents-a-barrel increase in the posted price of crude oil. Other benefits increased national tax takes by three billion dol-lars in 1971. At the same level of production as in 1971, the deal guaranteed an additional two billion dollars annually for countries until 1975.

It was the most expensive agreement companies ever had nego-tiated with producing countries. But that did not end OPEC de-mands. A few months later, in July 1971, OPEC decided that member states should participate in equities of companies. OPEC won that battle in January 1973, and the leapfrogging began again. Both Libya and Iran, acting separately, pressed for better deals. Iran, as mentioned earlier, took full control of its oil industry August 3, 1973. Kuwait, restive and feeling that it was falling behind in squeez-ing oil companies, called for a reopening of that participation deal of January 1973. Libya took note of Iran's victory and became even more determined to tighten the screw which it had on oil compa-nies. In those early days of August 1973, it appeared that Libya might force a showdown with all the companies operating within its borders. There were threats of a production shutdown which would introduce chaos into international oil markets that already were in short supply.

In international oil, trails lead around the world. A happenstance in Iran might be revealed through a chance meeting with an oil executive in Paris while a casual remark by another executive in London may be amplified by a Kuwaiti minister in his office. The denouement to the scenario which opened with that pipeline break in Syria in May 1970, has one scene set in the National Hotel in Moscow, that ancient hostelry located across a broad square from the red brick walls surrounding the Kremlin.

After finding me in their negotiating session with the Soviet Oil Ministry in Moscow, the Dresser Industries' officials involved had invited me to dinner with them.

"Maybe you can tell us what is going on," Ned Fowler, chief engineer for the firm's Oil Products Division, quipped. It was Igor A. Neudachin, director of Dresser's Soviet operations, who suggested the National restaurant as one of the best places in town for dining. In Moscow, "best" is a relative term, but Neudachin, a big, breezy extrovert of White Russian expatriate extraction is the sort of person who knows where the wine cellar is located, who is the chef in the kitchen, and how to stimulate a waiter on the floor.

The National's dining room has the faded elegance of Czarist days, cut glass chandeliers, tassles on lampshades, heavy drapes and everything which might have contributed to luxury in 1914, without much refurbishment since then. We were enjoying the shaslik when Neudachin nudged me, gestured over his shoulder.

"Isn't that Hammer in the corner?"

I glanced around. Sure enough. Dr. Armand Hammer, the medically-trained businessman who is chairman and president of Occidental Petroleum Corp., was dining at a corner table with his wife and an aide. Dr. Hammer has had a legendary career. He went to Russia fresh from medical school during a famine and typhus epidemic, moved by a do-gooding spirit. His success in handling wheat and medical shipments to Russia brought him into business. He represented 37 American companies, manufactured all of Russia's lead pencils. In cash-short Russia, he took profits in art.

After a ten-year career in Russia, he returned to the U.S. with a fabulous art and antique collection, which provided the basis for launching the Hammer Galleries, New York. He was successfully involved with several American companies, capping his career by taking leadership of a small oil company and building it into giant Occidental. With his U.S.S.R. background, he is in a unique position in this period of detente when Soviets seek American technology to assist industries.

I had met Dr. Hammer several times before, so I went over to his table to say hello, and found him in an affable mood. He has that ability to make an acquaintance feel important and now he said: "I'm not seeing any newspapermen in Moscow, but I'll make an exception in your case, Ray. Come see me at nine tomorrow morning in my suite."

With Libya's oil production on the verge of being shut down, I knew what I wanted to ask him the next morning. Occidental is one of the key petroleum producers in that country, and Dr. Hammer carries the company's operations in his head. If Libya's daily production of two million plus barrels per day were lost, petroleum shortages would be appearing in Western Europe in a matter of days. And the feedback would be extending to the United States shortly after.

Next morning Dr. Hammer was more interested in talking about the new Trade Center being built in Moscow by Occidental and the $8 billion chemical deal his company has concluded with the Soviet Union. Since Occidental gets all of its oil outside the North American continent in Libya, the sequence of his conversation told me things about Libya though he hadn't mentioned the country. There wasn't going to be any shutdown in Libya involving Occidental.

"And what about Libya?" I finally asked.

He confirmed what I had already deduced. An agreement between Occidental and the Libya government was "imminent." The government would take over 51 percent of the company's equity in the Libya operation in return for compensation fixed at the book value of the assets. In return Occidental would obtain a guarantee of oil from the government's share which would be produced.

That story, carried over the Dow Jones ticker and in the Wall Street Journal, prompted indignant denials from other oil companies operating in Libya. Even Occidental's public-relations department in Los Angeles issued a statement disavowing what the chairman had said. The next weekend, the Libyan government announced in

Tripoli that an agreement had been concluded with Occidental. It followed the lines of the settlement which Dr. Hammer had sketched for me in Moscow.

The tale of that Tapline pipeline rupture was about finished, its ramifications stretching around the world. Nobody can say, of course, whether or not the domino effect of company capitulations would or would not have occurred as they did if there had not been that rupture. The pipeline break certainly provided producing countries with the negotiating ammunition for those capitulations. The domino effect came sooner than it might have if the break had not occurred.

Meanwhile, as Arab nations were chipping away at the facade of the companies, the struggle was followed uneasily from afar by Israel, another Middle Eastern country which has virtually no oil of its own, yet which can affect every Arab oil producer in a near suicidal manner.

XIV

The Israelis

Masada's barren mountain towers above the Dead Sea in Israel like the fortress it once was after the destruction of Jerusalem in A.D. 70. Besieged Jewish Zealots here battled an overwhelming Roman force for three years before being slaughtered almost to the last man. The handful of survivors then committed suicide rather than surrender to the alien army.

The summit is reached by a serpentine path up a steep precipice when approaching from the Judean Hills. This is the hard way since the cable-car system was erected on the Dead Sea side. The footpath twists and turns in hairpin swings, up and up above a landscape akin to that of the Grand Canyon. It is easy to see why Masada is an Israeli national monument. Today it is revered as a testament to Jewish fighting spirit, a natural shrine where man seems insignificant in that tremendous sweep of rocky ridges and distant peaks.

The Zealots are long gone, but Masada remains with its ruins of King Herod's palace on a stone prow that points toward the Dead Sea—an unconquerable fortress, for the Romans conquered only the bodies of those Zealots, not their spirits. This is the way the Israeli see the mountain, and their own land of Israel. And this sense of existing in a fortress permeates the life of nearly every Jew who now inhabits this promised land of Moses. Every able-bodied man under fifty-five is in the army, along with many women under thirty-four. The defense forces drain 42 percent of budget spending.

"Peace is far away," said one moustached Israeli lawyer and reserve army officer who insisted on wearing a business suit even on the long climb up the mountain. Standing on a terrace of King Herod's palace, he fatalistically added, "So be it. The next war will last thirty days instead of eighteen. Results will be the same."

"But losses will be much higher for us this time," said his youthful, Levi's-clad wife, who looked more like a daughter than a spouse.

"For them, too," he said. He stared across the broad expanse of the Dead Sea. In Jordan, on the other side, the purple mountains of Moab, from which Moses first viewed the Promised Land, rose in folds into the clear blue sky.

The woman's Levi's fit the Israeli character better than did his suit jacket. Normally, the Israeli, especially the Sabra, or Israel-born Jew, is about the most casually dressed person on earth. He is proud of his proletarianism, unwilling to admit that a tie or jacket adds anything to a man's prestige. He has a chip-on-shoulder chutzpah and a willingness to upset protocol if it gets in the way of the job. In manner, he is much like the outdoorsman in the American West. Indeed, his own country resembles the American West—the Arizona–like Negev in the south, the California–like coast with its orange groves, the New Mexico–like mountains around Safad, and the western reservoir aspect of the Sea of Galilee or Kinneret.

Significantly, the word "sabra" comes from the fruit of the prickly pear cactus. The true Sabra probably does seem to have a prickly disposition and exterior, though, like the fruit, he may have a soft heart.

Immigrants often retain the customs and prejudices of the lands from which they came. This is why it is difficult to find a master pattern for the Israeli character, though the Sabra undoubtedly comes closest to filling the role. This particular lawyer atop Masada came from Germany and he still had some of the Prussian formality that has almost disappeared in Germany itself.

When I mentioned that Israel had to find its place in the Middle East, he stiffly retorted, "Israel is a Western nation." Geographically and ethnically, however, Israel is also a Middle Eastern nation, a part of the area whether or not Arabs are willing to admit it, and whether or not Jews are willing to accept the "Middle Eastern" designation. Sometimes the East European or American-born Jew bridles at that suggestion. This doesn't change geographic or ethnic compositions.

All other states in the Middle East contain a Moslem majority, except for Lebanon, where the ratio is half Christian, half Moslem. Israel is a Jewish state, the national home for 2.8 million Jews plus any others who wish to leave the diaspora (the Jewish world outside Israel) for aliyah (moving to Israel). Yet 460,000 of its population is non-Jewish, mainly Arab Moslem. Moreover, two-thirds of the Jewish population was born in the Middle East or North Africa, with cultures reflecting the lands of birthplaces, a percentage that is rising steadily as time moves on and the dominant Eastern Europeans are growing old and disappearing. These figures, based on 1972 population counts, do not include the more than one million Arabs who are in the occupied territories seized in the 1967 war. Politically, they are nonpersons, and are likely to remain so. The 1973 war added no additional Arabs in any numbers.

Were Israel to annex these occupied territories, the Arab population within its borders would total a million and a half versus 2.8 million Jews. But Arabs are much more prolific than are Jews. In another few decades, the Arab population in Israel then would outnumber the Jewish. This is one reason why Israelis fear anything that resembles a multiracial state. If Palestinians returned to create a state involving both Arabs and Jews, the Jews would be already outnumbered.

How many Palestinians would return if it meant living under Israel rule? Not more than a few hundred thousand, and probably much less. Compensation might satisfy the bulk of them. And it is difficult to imagine Israel taking any more than a token number back.

The Jewish population is far from being homogeneous, as it is. This little nation is truly a melting pot of peoples. There are dark-skinned Yemeni, yellow-complexioned peoples from Cochin China, and blonds who may have come from Poland or elsewhere in northern Europe. In all, more than eighty countries are represented in the makeup of the Jews of Israel. Israel really is a mosaic of peoples held together mainly by the common bond of religion, and the realization that they must live together as a nation or die separately.

The vigor and drive of the Eastern Europeans helped create Israel. Their socialistic ideas permeate the structure of the state, and they have provided most of the political leadership throughout the history of modern Israel. Many of them also are unable to understand the Arab or his mentality. Some cannot even understand the Oriental Jews, those Israelis who have cultures and backgrounds more akin to the Arab than to the European.

Thus, the chasm that exists between Arabs and Israelis is due not only to religion but also to the clash of Old World ideas and thinking with the culture of the Arab. Add land claims to this clash and the problems of peace sometimes seem insurmountable. Yet peace some day must come to the area, for all wars fade and die under the weight of time—the sooner the better, insofar as America's oil problems are concerned, for Israel's well-being too.

The problems of this peace, of course, are two-sided. Once in the Basman Palace, in Amman, Jordan, I encountered Wasfi Tell in the anteroom before King Hussein's office, shortly before he was to become prime minister, only to be shot dead by an assassin on the steps of the Sheraton Hotel in Cairo.

Since a country like Jordan needed so much development, why, I asked him, could not the technology and the drive of the Israeli be wedded with the manpower of the Arabs to transform the whole region into a prosperous land? He had a sour expression on his face as he answered, "If such a situation like that developed and there was freedom of movement and capital, the Jews would end up by owning Amman."

Israelis have the most finely honed humor in all of the Middle East. They can even joke about something as serious as peace. Once in a discussion at the home of a professor of political science at Tel Aviv University, the question was asked: Does Israel really want peace if peace means surrender of the occupied territories?

"Sure we want peace," said a youthful professor who had been critical of both Arabs and his own government. "We want a piece of Syria, a piece of Egypt, and a piece of Lebanon."

That same evening I heard another story which says something about both the characters of the Arab and of the Israeli.

Ahmed, the Arab, was sleeping in the town square. Youngsters were playing in the neighborhood, making a racket. Ahmed was awakened, and he knew he wouldn't be able to sleep again with those kids around. So he concocted a story.

"Listen," he said to the children. "Sheikh Ibrahim is giving away money on the other side of town."

Happily the kids scampered away in the direction of the Sheikh's house. Ahmed lay down on the pavement and closed his eyes. He was just drifting off to sleep when the idea penetrated into his mind. Sheikh Ibrahim is giving away money!

He sat up, suddenly, shook the sleep from his eyes. "If the Sheikh is giving away money, I had better hurry over there and get some, myself."

Arabs often do seem to believe their own propaganda about Israelis, and the Israelis, in turn, often view the Arab as a simple-minded person who must be treated like either a child, or a wild fanatic who should be handled like a mad dog. The stories one hears about Arabs, of course, are very similar to the tales one hears in America's Deep South concerning the Negro.

We happened to be dining in an Israeli home in Tel Aviv the day news broke that West Germany had freed the Arab killers in the Olympic Games massacre at Munich. Freedom was granted under duress after Arab guerrillas hijacked a Lufthansa commercial airliner, which had just left Beirut, then threatened to blow up the plane and passengers unless their jailed compatriots in Germany were freed. West Germany capitulated to the demands.

There was a moment of stunned silence at the dining table that night when the news was revealed. The hostess, a pleasant, attractive girl of culture and refinement, who has compassion for her fellow men and a strong interest in justice, exploded. "Why did the Germans give in to them? Those passengers on the plane were only Arabs!"

Every Israeli seems to be carrying a rifle. Drive an automobile around the country and you find yourself stopping and starting in jerky sequence, picking up and dropping off hitchhikers in uniform, both male and female. They seem to be at every crossroad corner in droves, going on leave from camp, departing for camp.

"The army is part of our lives," said Dan Margalit, one of the most knowledgeable political writers in Israel. I met him one night in Tel Aviv on the day of Lag B'Omer, the only time in the seven-weeks' period between Passover and Shavuot when the Orthodox Jew may be married. Mrs. Margalit was attending two wedding receptions on the same night, and Dan was baby-sitting. She outranks her husband in the reserves, holding the rank of sergeant.

The Israel Defense Forces consist of a nucleus of commissioned and noncommissioned regular officers, a contingent called up for national military service, and a reserve. Men under twenty-nine and women under twenty-six are called for regular service of up to thirty months for men and twenty months for women. Physicians, both male and female, may be called up until the age of thirty-four. Married women, mothers and pregnant women are exempted, as are Arabs (which some Israeli Arabs regard as one expression of discrimination against them).

After serving their hitches, the draftees move into the reserves, men until fifty-five, women until thirty-four. Men train annually for thirty-one consecutive days until they reach the age of forty; from then until fifty-five, the period is fourteen days annually.

Israel can marshal a formidable army in hours, as it did on October 6, 1973 when war again started. I was in the Jordan Valley that afternoon, drove that night through partial blackouts from Jerusalem to Tel Aviv. Hundreds of youths hitchhiked on highways, heading for call-up points. There was no doubt that Israel was caught by surprise, yet the system worked.

"War is a question of technology and good morale," said Margolit. "We have both."

It is a costly matter for Israel. Defense accounts for one-fourth of the country's total gross national product. Before the 1967 Six Days War it had been only 11 percent. By contrast, the United States, even at the height of the Vietnam War, was only spending 8 percent of its GNP on defense.

"We are now a laboratory for new weapons," said Moshe Sanbar, governor of Israel's Central Bank in Jerusalem. I found him very concerned with spending trends, worrying about Israel's ability to carry the heavy defense load unless it obtained substantial help from abroad, mainly from the United States.

Sitting in the coffee shop of the Parliament Building in the capital, I listened as Yad Jacoby, a member of parliament and Deputy Minister for Transport, expanded on the country's money problems.

"Half of Israel's defense budget is now carried by the United States, by the Jewish community making contributions and buying bonds, by U.S. federal government loans and grants, and by other such help," he said.

It is the extent of this help that angers Arabs. Everywhere an American travels through the Arab world today he hears criticism of the manner in which the United States is supporting Israel. Where would Israel be without that help? A good question, but like all hypothetical questions about what might have been, there is no answer for it. Do not underestimate the vitality and power of this little nation of 2.8 million Jews, which seems able successfully to defy 100 million Arabs. If Jews had received no help at all from outside, Israel might have managed to survive all by itself.

Israel certainly would be having a much rougher time, though, without American help. In addition to the external Arab problem, it also has disputes and conflicts within its own social structure. There are cleavages between immigrants and old settlers; between Westernized Jews from Europe and America on the one hand and the Oriental Jew from the Middle East and North Africa on the other; between the Zionist concept of a religious movement centered on Israel but encompassing world Jewry, and a narrower nationalism that focuses on Israel alone; between the religious and the secular; between socialism and the desires of free enterprisers to shake loose of the fetters imposed on the economy by the bureaucratic organizations of socialist pioneers; and between the haves and the have-nots.

The Oriental-Western Jew schism personifies some of the social cracks that exist. Israel's Oriental Jews are at the low end of the social and economic scale, often living as an inferior minority even though this social class now is in the majority. The children of Oriental Jewish parentage comprise 70 percent of the school classes when children start in school. Only 16 percent finish high school, and only 3 percent finish university. In some respects the Oriental Jew is the "black" of Israel.

"We have sixty to seventy thousand families who are living at a substandard level. Ninety percent come from Oriental communities," said Yitzchak Ben-Aharon, Secretary-General of Histadrut, the million-member labor-union organization that operates health and welfare services like a state within a state and controls 25 percent of the country's economic sector.

Before seeing Ben-Aharon I had been warned that "he hates newspapermen but likes publicity." I found him to be a friendly, outgoing man who looked much younger than his over-sixty-five. He had positive opinions and talked bluntly, especially when discussing the less fortunate among Israel's citizens.

Taxes in Israel are so high that the country is unable to raise the revenues needed for solving its welfare problems. "You can't move without paying taxes," Ben-Aharon said. "There are taxes on coffee, on tea, on the notebook you buy in a stationery store, on shoes, cigarettes, everything. We have the highest rate of taxation in the world."

He suggests that more, not less, help must come from outside if Israel is to absorb the Oriental Jews and others into its social structure without creating a segregated class of have-nots. Pinhas Saphir, Israel's finance minister, is a key man when it comes to raising funds abroad. He is a big bear of a man, with a hairless head, a gruff, aggressive manner, and an overpowering personality that is a liability when he appears before political audiences. Yet he is a consummate politician for all of that, a master at the negotiations that occur in the Israeli equivalents of smoke-filled rooms.

"Before, our people were only conscious of the shooting and the threats of the Arabs," said he. "Now we have easily defendable borders. So people are thinking more in terms of progress, of schooling for their children, of housing." I interviewed him just after he had met with officials of the Housing Ministry in the Finance Ministry's building in Jerusalem. I had noted the mournful faces of the housing people when they filed out, after having their budget slashed by the forceful Saphir. Just as forcefully, Saphir emphasized that outsiders should not misunderstand the significance of those cleavages that are apparent in Israeli society. When danger threatens, the Israeli have a habit of drawing together very quickly, he stressed.

The fighting spirit of the Israeli is evident in the Sinai, the bleak peninsula of rock and sand that was seized in the 1967 war and reaffirmed in that of 1973. Sharm el-Sheikh, at the southern end of the Sinai Peninsula, has become akin to a national shrine for Israelis. Arkia, the internal airline, and Egged Dan, the tour agency, operate frequent tours to it. New hotels offer accommodations. A new road from Eilat at the head of the Gulf of Aqaba makes it possible to drive to Sharm el-Sheikh from Eilat and back again in one day over a spectacular road. On some days one may encounter several hundred people at the point, and the bulk of them seem determined that Israel should retain the area.

The reason is that it overlooks the Straits of Tiran, the narrow waterway that permits passage between the Red Sea and the Gulf of Aqaba. Were that passage closed, Israel would have no southern port. Ships that now traverse the Red Sea and the Gulf of Aqaba would have to go all the way around southern Africa and into the Mediterranean to reach Israel.

Standing beside a spiked Egyptian gun, which was rusting in the bright sun, Benzion Spector, operator of a hotel at Sharm el-Sheikh, pointed toward the 1,500-yard-wide Straits of Tiran, where coral reefs squeeze the channel. He said, "The guns here of the Egyptians threatened to close our port of Eilat in 1967, and that was why the Six Days War started. How can we give this back to Egypt to let them threaten us again?"

Israelis do not seem to have any intention of giving up control of Sharm el-Sheikh. And Egypt insists it must be returned along with the rest of the Sinai if there is to be peace. This is the sort of situation that seems to predominate in the Middle East, the impasse that apparently has no solution.

The question of possible compensation for the Palestinian refugee is apt to bring up a counter question: What about compensation for Jewish refugees from Arab lands? Around 500,000 Jews migrated to Israel from Arab lands. Many of them were forced to abandon property and wealth when they departed. If the Arab refugee deserves compensation, does not the Jewish refugee, too, ask the Israeli?

Most of the 23,622 square miles of the Sinai is bleak desert and spectacularly beautiful mountains, especially in late afternoon when the slanting sun pours red and pink light onto the rocky cliffs and canyons of jumbled ranges. But except for a few settlements along the Mediterranean shore, oil camps on the Gulf of Suez, and Saint

Catherine's Monastery in the heart of the Sinai, only a few Bedouins occupy the whole peninsula. It does not seem as if it is worth extending the Israeli-Arab confrontation. Israelis, however, like the distance the Sinai places between Israel and Egyptian armies. This, they claim, is the key reason why they want to cling to some of that territory.

In Tel Aviv, one hears, too, that the Sinai is proving rich in minerals, that it could have solid economic worth for a nation like Israel, which possesses few natural resources outside of its hard-working people. Already Israel is pumping oil from the fields that an Italian company, Ente Nazionale Idurocarbi, had been exploiting for Egypt until the June 1967 war.

In 1973, a little over 100,000 barrels a day were being extracted from the wells around Abu Rodeis, the compound of Italian-style villas and barracks that lies beside a limpid Suez Gulf. On the far side, at night, the flares of burning gas at Egypt's El Morgan field light the sky.

Is oil a factor behind Israel's reluctance to surrender the Sinai? "Peace certainly means more to us than oil," said one Israeli government official, when asked that question. Then he added, "But as long as peace eludes us, we might as well keep pumping this oil."

In the guest book at Saint Catherine's Monastery, where Greek Orthodox monks have one of the best collections of rare books in the world, there is a comment written by Israel's Colonel Abraham Brent immediately after the Six Days War in 1967. It says, "The first thing I did when the war was over was to come here with my friends. I pray that we will keep this wonderful place in our hands."

Arabs feel that Colonel Brent's sentiments are widely shared in Israel. Now Arabs view their oil as a possible instrument for promoting a favorable solution (for them) of the Israeli-Arab confrontation. To them, the energy gap in the industrial world seems to be a fortunate occurrence that immeasurably strengthens an Arab position that militarily is ultra-weak. Israelis have the best army and air force, man-for-man, in the world. If another war developed within the foreseeable future, Israel might win again at heavy cost. But more and more it is becoming apparent that American military technology and pecuniary largess are big factors in enabling Israel to maintain itself as a fortress nation.

It is not only Arab sources that feel that if the United States is made to experience acute oil shortages, it may be forced to seek compromises in the Middle East. At the March 1973 Europe-America Conference in Amsterdam, Neville Brown, a political scientist with Britain's Royal United Services Institute for Defence, gave this warning:

A threat to European and indeed North American security is that by the 1980's the Arab World will be more willing and able to apply crippling oil sanctions against the West in order to make the West in its turn oblige Israel to relinquish East Jerusalem and other areas occupied in 1967. For by 1985, the non-communist world is authoritatively expected to be consuming two and a half times as much oil as it was in 1969, in spite of the advent of nuclear power in electricity generation and other things.

The Yom Kippur War speeded up that timetable and before 1973 ended some West European long-time allies of the United States were suggesting that America should reconsider its Mideast policies because of the energy situation.

This possibility frightens Israelis, too. Nearly everything said or written about the world's energy gap is being relayed to Tel Aviv and to Jerusalem, today. There it is assiduously studied by the military, by intelligence, by the Foreign Ministry, and by other sections of the government. Like it or not, Israel sees the possibility of its being drawn into this oil-energy question, with its political positions at stake. Heretofore, it usually has possessed the initiative in any crisis affecting its national health.

Now Israel faces a situation where the initiative may lie elsewhere, in the capitals of the Arab world, where immense financial and economic power will be resting in the years ahead. This is a sort of power where Israel's admitted overwhelming military superiority means little, which is an unhappy prospect for Israel. It is for the United States, too, for America faces shortages of fuels, the necessity for imports of energy supplies from the volatile Middle East, soaring fuel prices, and a balance-of-payments drain for oil that may be in the fifteen-to-twenty-billion-dollar-a-year range.

XV

The Road Ahead

A Chevrolet Impala eats up the miles on the asphalted road between
Jidda and Riyadh, the speedometer registering a smooth eighty miles
an hour on the flat desert stretches. Rock and sand flash by in a
tawny blur, empty space stretching to hazy horizons beyond the
immediate moving foreground. It is a road for traveling, for making
time, for sinking into deep thought while reflecting about the vast
changes underway in the Middle East today.

For the Arab and the Iranian, the road indeed is paved with gold.
Iran is becoming a bigger market for goods and services than is India
with its 550 million population. Kuwait has so much money in its
treasury that it no longer sees any reason for pushing oil production
to maximum capacity. Saudi Arabia is becoming a monetary power
with so much financial strength that it has become a discussion topic
at meetings of the International Monetary Fund, the agency that
helps coordinate the world's money system.

By 1980, the oil nations of the Middle East may be earning over
sixty billion dollars a year from their petroleum. (James E. Akins,
ex-director of the United States Department of State's Office of
Fuels and Energy, estimates that Persian Gulf producers alone may
be earning fifty-eight billion dollars annually from oil in 1980.) These
nations seem to be spending their money wisely, for development
projects that benefit the people. There is an obvious drive everywhere to upgrade educational levels, to alleviate poverty, to raise
women from the purdah into which they had been cast, to build the

infrastructures of modern states, and to diversify economies. Unfortunately, a lot of money goes into defense, too. But Arabs are just as worried about "secure borders" as is Israel and now some states have the funds for expensive weaponry and for the training systems necessary for creating the manpower to wield that weaponry.

The financial overspill of massive spending programs splashes onto non-oil nations in the Middle East such as Lebanon. It washes into exporting factories in Europe, America, and Japan. Within the span of an hour in a Jidda Hotel, I recently encountered sales and market research representatives of General Telephone and Electronics Corporation, New York (communications equipment); Allis-Chalmers Corporation, Milwaukee (farm and construction equipment); McCracken Concrete Pipe Machinery Company, Sioux City, Iowa (pipe machinery); Arpol Petroleum Company, New York (precision oils); Besser Overseas Corporation, Alpena, Michigan (concrete block machinery); United States Steel Corporation, Pittsburgh Pennsylvania (steel); Reed International Sales Company, Erie, Pennsylvania (hand tools); KDT, Inc., Huntington Woods, Michigan (export agent); The Architects Collaborative, Cambridge, Massachusetts (architectural consultants); and Metcalf and Eddy, Inc., Boston (engineering consultants).

All were looking for business. Some already were getting it.

"There is business to be had for American companies all through this area," said Frank Ladue, President of McCracken Concrete, a jolly salesman with chin whiskers who had the unusual distinction of having been bitten by a camel in Abu Dhabi. He was busy selling complete concrete pipe factories that would be manufactured in Sioux City, Iowa, and put together in the Middle East.

Ross P. Wright, President of Reed International, said enthusiastically, "Over the next five to ten years, this area will be the most booming in the world."

America's exporters certainly have a job cut out for them. It is vital that the United States increase its exports of everything that it can over the next few years in order to prevent dollar devaluation after dollar devaluation. Peter G. Peterson, former United States Secretary of Commerce, provided some chilling statistics concerning U.S. trade.

"Based on estimates of U.S. energy demands, I am told that the projected U.S. trade deficit in energy could be in the range of $15 to $21 billion by 1980. The bulk of our imports will be in oil, and most of it will come from a handful of Middle Eastern and African countries," he said.

That balance-of-payments problem came up for discussion again at a European Investment Seminar sponsored in Paris in early February 1973 by Burnham & Company, Inc., New York brokerage house. Thornton F. Bradshaw, President, Atlantic Richfield Company, Los Angeles, presented the oil-industry picture in a comprehensive paper and brought up the question of how America is to pay for its oil. During a break in the sessions, I sat across from him at a table in the Coffee Shop of the Inter-Continental Hotel, talking oil and finance.

"This trade deficit for oil poses very real monetary problems for the United States," he said. There was a mournful note in his voice when he added, "I have not seen any solutions set forth anywhere. I have not even heard the promise of any solutions."

Evidently America has a trade problem on its hands that should be drawing attention along with the energy gap. The United States needs to participate in a liberalization of world trade to increase its export earnings. Only in that way can it pay for the petroleum it must import. This is one of those areas where trade protectionism will do no good, and perhaps much harm, for the United States must import oil or shut down factories and curtail motor-vehicle traffic.

Of course, there are other things the United States can do, too, to alleviate the problems connected with its energy gap. Fortunately, people are becoming aware of problems, the first step toward finding solutions in any democracy. With appropriate action, hopefully the worst fears of energy pessimists may not be realized.

America's energy crisis has spawned a great debate that rages in the media, at intellectual seminars, and in business conferences around the country. Is the situation really serious? Are oil companies creating a myth to bamboozle the public or is the United States really short of energy? If so, what should be done about it?

Some sources blame the international oil companies for everything, as if corporations may have enjoyed being kicked around by producer nations. Oil company profits were rising in 1973 which was considered immoral by some people, as if companies aren't supposed to show a profit in a free-enterprise economy.

In any area with as many ramifications as energy, there are bound to be as many answers to questions as there are axes to grind. One school of thought contends that there really is no energy crisis. We are seeing the consequences of misgovernment, the result of delay in introducing a much-needed energy policy. Political complexions of proponents of this school are evident.

That other camp has focused sights on the international oil companies—the Exxons, the Texacos, the Mobiloils, and such. To these business detractors the situation represents a devious plot on the part of companies to boost oil prices and to further their own selfish interests. This group wants nationalized oil companies to supersede the private, free-enterprise internationals, with major questions in world oil decided on a government-to-government basis. Many people in this camp don't try to hide their socialist leanings and anti-free enterprise prejudices, because they can't.

A related school contends that international oil companies should be broken up, with production and refining operations separated. In July 1973, two anti-trust cases were opened by the states of Florida and Connecticut aiming at just this, while the Federal Trade Commission was readying federal action. Obviously, if any laws have been broken, violators should be punished. However, from evidence presented so far one of the sins of the big companies seems to be their bigness, in the FTC's view. Yet one wonders how companies would manage against Middle East producing countries in negotiations were firms fragmented into many small concerns. This wouldn't produce one more drop of crude in a tight situation, and it might result in less oil. Moreover, since downstream operations aren't paying their way now, how will they pay their way if divested from the production end? The answer is obvious, of course, through much higher prices at the merchandising end. Is this the aim of these suits?

The anti-Israel faction marshals data to show that the situation is so serious that America and Europe must support Arab positions vis-à-vis Israel for protection of oil supplies. The pro-Israeli faction counters with claims that America and Europe must reduce their dependence upon Arab lands lest the Atlantic Alliance find itself hostage to the Middle East.

Environmentalists have their own answers. If people left their automobiles in their garages most of the time, kept home furnaces at sixty degrees Fahrenheit in winter, scrapped air-conditioners, and reduced usage of home appliances, energy use would decline drastically. There would be no shortages and people would have a cleaner atmosphere. They also would have a lower standard of living, except for the clean air. This possibility is seldom mentioned.

So it goes. A "fact" may be any statistic you espouse. All other data is treated as "fanciful."

It may seem difficult to term the so-called energy gap as a myth when some gasoline pumps run short and fuel-oil distributors have no oil for homes and factories. Incredibly, there are people who say such spot shortages are "temporary situations unrelated to the over-all energy question." They do not see that when supply and demand are finely balanced, even an apparently "temporary" slight disturbance may cause price rises and/or shortages.

Today, it does not take much to upset oil markets. Minor troubles are only the surface indications of the broader problem underneath —the energy gap, the energy crisis, energy shortages, or whatever else one elects to term it.

Actually, much of the industrial world has long had energy shortages, if by shortages one means the inability to produce enough at home for needs. Japan depends upon oil for 85 percent of its energy and obtains 100 percent from imports. Western Europe produces about 3 percent of its petroleum requirements and imports the remainder. Even after North Sea oil arrives, Western Europe still will be importing 85 percent of its petroleum requirements.

The new element in oil is this: the United States, long the world's largest producer, has lost its self-sufficiency. Demand outruns its ability to produce. In 1980, America will be forced to import about half its oil, the bulk of the imports from the Middle East.

This new element is having repercussions upon the world. Heretofore, America's spare-oil capacity served as a reserve for Europe when the latter had supply problems. Now the spare capacity is gone. This tightens the pinch upon Europe, and to a lesser extent upon Japan, which, in turn, increases competition for oil. The Arab positions are strengthened as they bargain for higher prices through the Organization of Petroleum Exporting Countries. As an importer, the United States faces those higher prices too.

The situation is accentuated because all crudes are not alike. There are variations in the weight of oils, in sulfur contents, in other chemical compositions.

Even two adjacent fields may produce different types of oil. International oil companies serve as blenders of oil to meet requirements of customers. The Shell Group, for instance, reports that to satisfy Britain's needs the group must import twelve different kinds of oil from eight different countries. Oils are blended in the way that a party host may mix different types of alcohol and other beverages to produce cocktails for guests. The analogy may be oversimplified, but it fits.

Today, low sulfur crudes are popular because of increasing concern about pollution. Thus, oil from Abu Dhabi and other low-sulfur crude centers is in strong demand. High-sulfur crudes sell less even in a sellers' market.

Oil from African fields and the forthcoming North Sea oil are lighter than Middle Eastern crudes. This needs less refining to produce naphtha, the raw material for gasoline. Now more of this type of oil is becoming available. The economic effect of this is being experienced in Europe. As America takes more of the heavy crudes from the Middle East, Europe finds itself receiving a larger percent of light oils than in the past, and gasoline is thus more readily available. Heavy crudes for power plants, however, are in shorter supply.

Thus, coupled with the energy gap is the question of how to balance the supplies that are available. International oil companies with their integrated systems are well equipped to handle the blending-oil distribution role. In fact, this is one of the arguments they raise for their existence, as governments intervene more and more in markets. Companies take crude from Canada for sale in the United States, dispatch other crudes to Canada for its special needs, and mix Texas oil with that of Libya or perhaps blend California oil with that of Peru. Whether or not nationalized oil companies could do this as cheaply and efficiently as private companies in the business for decades is another story.

247

The energy gap is confined to specific countries. There is no world shortage of oil at this time, and there is unlikely to be any in this century. Saudi Arabia alone has reserves of over 150 billion barrels of proven oil as of early 1973, and more oil is being discovered all the time. The total world oil reserves at the end of 1972 were estimated at 672.7 billion barrels by British Petroleum Company, Ltd. Of that, a very conservative 52 percent, or 355.3 billion barrels, were estimated to be in the Middle East. North Africa held another 7 percent, or 47 billion barrels, which gives Arabs and Iranians nearly two-thirds of the world's known oil.

But oil in the ground is far different from gasoline in the tank of your automobile. Unfortunately, petroleum reserves are unevenly distributed—60 percent in the Middle East and North Africa, and only 40 percent in the United States, Western Europe, the Soviet Union, and the rest of the world. Thus nature forces the world to deal with the Middle East when it comes to oil.

By why oil, you may ask. What about other sources of power? Unfortunately, nuclear power cannot be injected into the gasoline tank of your automobile. Yet there are other sources of power for certain things.

In 1973, America's energy was being supplied as indicated in this table from the Chase Manhattan Bank, New York:

	1973	1972	Percent change
(Million-Barrels Daily-Oil Equivalent)			
Oil	17.3	16.3	+6
Natural gas	10.9	10.9	–
Coal	6.5	6.4	+2
Water Power	1.4	1.4	–
Nuclear Power	.6	.3	+100
Total	36.7	35.3	+4

Starting from the bottom of the table and working up, you may obtain an idea of energy patterns. Nuclear power is arriving much slower than its proponents claimed it would a decade ago. It is expensive, and ecologists battle hard to prevent the location of nuclear power plants in their neighborhoods. Oil overall has been so much less expensive that electrical-power companies have preferred it as an energy source.

Water power can provide little energy. It can be expanded only marginally.

Coal, like nuclear power, is hamstrung by the ecologists, who oppose strip mining, which is far and away the cheapest method of mining coal. High sulfur–content coal is shunned, and smoke-pollution regulations prompt industrial users to switch to oil. As a result, coal use represented only 18 percent of total energy consumption in 1972, an estimated 17.7 percent in 1973.

As for natural gas? Ah, there is a story of bungling government intervention regardless of which party happened to be in power. For years, the Federal Power Commission and various state regulatory agencies have fixed the prices of natural gas. For years, those prices have been held so low that it was uneconomical for oil and gas companies to develop new sources, and this at a time when pollution controls were regulating coal away from certain markets.

When the demand for gas climbed, the supply could not keep pace. Thus we are now experiencing a shortage of natural gas. This is a classic case of government interference in a market, allegedly to protect the consumer, with opposite results; of course, no politician or bureaucrat ever will admit error. It is far easier to blame those "wicked oil companies" for conniving against the poor consumer. So when you hear some government man damning companies in any industry, look closely before giving him your support. He may be right. He also may be laying a smoke screen around himself.

And finally, what about oil?

Every natural resource is a depleting asset, and oil is no exception. This is why federal law permits an oil-depletion allowance for taxes. Perhaps the United States is not running out of oil; it simply is running short of readily available oil of its own. In a way, the United States is like New York City, which is adjacent to the Atlantic Ocean yet experiences water shortages. There is a lot of water in the ocean, but it is not in usable form.

There may be great new oil fields to be discovered in America's offshore holdings and in Alaska. America has passed its production peak, unless such fields are discovered and developed, or unless technology can find a way to unlock the oil contained in the shale deposits of Colorado.

With America's oil there are numerous "ifs" and "unlesses." Moreover, history shows that it takes five-to-seven-years' lead time before any new oil discoveries can be brought into production. Thus, there is little hope for filling any production gaps with American oil through the rest of this decade.

Meanwhile, the demand steadily rises, and imports will have to rise too. In 1971, the United States imported 3.8 million barrels of oil daily, with the figure rising to 4.6 million barrels daily in 1972. Estimate is that the 1973 figure was around 6 million barrels a day, or to about 35 percent of the total supply of oil. By the end of this decade, the figure may be 12 to 15 million barrels a day or over, for half of America's oil consumption.

Perhaps one can criticize the federal government for tardiness in developing an energy policy. The Federal Power Commission's support of cheap natural gas at the expense of supplies was perhaps somewhat shortsighted. Perhaps the international oil companies can be blamed, too, for underestimating consumption. Yet there are no real villains in this situation.

For years, consumer groups lobbied for low gas prices, as was their right. Even today certain groups still press for lower prices on oil and gas, as if the supplies will flow indefinitely. And government officials cannot be blamed for having listened to consumers in the past. This is the democratic way of doing things. If governments hadn't listened, some people now in office might not be there, today. Looking ahead, though, it appears that consumers might be served better with assured supplies at somewhat higher prices than with erratic or no supplies at low prices. Because natural-gas prices have been so distorted by law under regulations, increases are likely to appear shocking at the consumer level. In 1972, for instance, natural gas that sold for 26 cents per thousand cubic feet at the well head in Texas sold for 46 cents in New York. But Algerian gas may sell for $1 in the United States and Siberian gas may be even higher.

Environmentalists are something else. A clean environment certainly is desirable. Yet it does seem as if the rush to purge the industrial sins of decades has led to unrealistic ecological standards. Today, energy needs are on a collision course with environmental desires. It is ironic that in early 1973 when numerous cities around the country experienced energy shortages, the United States Supreme Court upheld the protests of ecologists against the construction of a pipeline across Alaska to tap Prudhoe Bay oil.

Already the development of Alaskan oil has been delayed for several years because of the environmental collision. And the importance of Alaskan oil should not be overstressed. With luck, production might get underway there in 1976. Yet its output, perhaps two million barrels a day by present reckoning, will merely make up expected declines in production on the American mainland. The need for imports will be great with Alaska, and greater without.

No oil company could have anticipated such strong ecological opposition when petroleum-consumption trends were being evaluated years ago. It hardly seems fair to hamstring the expansion plans of companies, then to blame them for failing to expand fast enough to meet the situation.

Today there are 250 oil refineries operating in the United States and nearly all of them run at or close to capacity, provided that adequate crude can be supplied to them. Oil-industry sources estimate that at least seven new large refineries will be needed on the United States East Coast by 1975 to meet demand for refined products.

They won't be there. Whenever a company suggested building a new refinery at a certain place, environmental objections usually were raised and that halted things before they could even start. Again, should oil companies be blamed if there are shortages of refined products on the Eastern Seaboard in 1975 and thereafter?

The international petroleum transport industry is shifting to the VLCCs, the giant supertankers of 200,000 deadweight tons and up. But there is not a single port in the United States equipped to handle such a ship. When a suggestion was raised that such a port be built off New Jersey, the ecological storm reached all the way to the state legislature and the governor's mansion. The United States Corps of Engineers' plan called for construction of a tanker terminal so far from shore that visiting ships would have been beyond the horizon of people on beaches. Oil would reach the shore through pipelines laid on the ocean floor.

Opponents of the port rushed a bill into the state legislature to prohibit any deep-water petroleum port adjacent to the New Jersey coast, despite the fact that one hundred such ports have been built in the United Kingdom, the European continent, Japan, the Middle East, and elsewhere in the last five years. The United States is one of the few industrial nations that does not have even one super port for these giant tankers. Yet the ten billion barrels of oil already carried safely in giant tankers would seem to indicate that they present less of a pollution problem than do much smaller tankers. Texaco, Inc., for instance, estimates that it took sixty-seven trips by tankers of 23,000 to 101,000 tons deadweight to service its Eagle Point, New Jersey, refinery in 1972. The same job could have been done with only seventeen trips by tankers of 250,000 deadweight tons each.

The American people will have to come to a decision. If energy needs and ecology clash, compromises may be in order unless people are willing to drop their living standards markedly. Certainly, if the majority of American citizens prefer the horse and buggy to the automobile, the whisk broom to the vacuum cleaner, and a chilly home to a cheery, warm one, then they have ways of making their wishes known to representatives in Congress and in state legislatures. Ecology would win.

But do the people really want this triumph? One suspects not when the price is realized. One suspects, too, that some of the ecological arguments are based on the fact that oil refineries, pipelines and concessions are all right just so they are adjacent to somebody else. It is akin to the argument we used to hear in racial matters: Integration is all right until the black man moves next door.

All life is a compromise. It is likely that citizens will have to compromise on ecological-energy questions. Nobody wants companies to pollute land and atmosphere. Nevertheless, denying them the right to expand is unrealistic when their business concerns America's industries, your job, and your living standards. It might be better to allow them to operate, but with strict provisions that force the polluter to pay so much that he adopts every safeguard to prevent polluting.

It is too late to think that measures introduced tomorrow may eliminate the energy shortage, though they might alleviate it. Here every possible avenue should be explored so as to reduce imports to the lowest possible level. The following steps would be helpful:

1. Increase incentives for the discovery of new oil and gas reserves. Permitting higher natural-gas prices is one example.
2. Expedite development of America's potential offshore areas.
3. Follow flexible oil import policies.
4. Use government funds to expand research into alternative fuels—geothermal, nuclear, shale oil, etc.
5. Resolve environmental conflicts, perhaps through a new government agency that might referee disputes.
6. Encourage greater coal use, since America does have enough reserves for about 400 years' consumption.
7. Encourage conservation, using the tax system if necessary. For instance, a second automobile in a family might be taxed much higher than the first; the vehicle tax might be graduated with rates rising sharply with engine horsepower.

Ezra Derek, head of Britain's National Coal Board, has said that 50 percent of all energy is wasted either by industry or by individuals. So this area alone provides opportunities for effecting savings that could improve overall supplies through the lessening of demand.

As an example, better insulation could be employed in buildings to conserve heat in winter and to make air-conditioning more efficient in summer. There could be more use of buses and trains for surface transport, and less advertising to encourage electricity or gasoline use. The tax system might force a trend toward smaller automobiles.

President Richard M. Nixon has advocated some measures along the above lines, but implementing a comprehensive energy policy will not be easy, and the United States is not going to escape its dependence upon the Middle East for imports over the next ten years, no matter what happens. In the early 1980s, however, the United States could reduce its import needs to ten million barrels a day instead of fifteen million barrels, through foresight and effort starting now.

And what about international cooperation among consumer nations to alleviate the problem? This is a question being asked to an increasing extent in numerous quarters.

"The problem of the future energy position of the Atlantic-Japanese complex of nations is one of the most important issues confronting not only each of them individually but also as a group," Walter J. Levy, New York–based international oil consultant, told the Europe-American Conference at Amsterdam in March 1973. The conference brought together several hundred political leaders and opinion makers from both sides of the Atlantic to discuss mutual problems and possible solutions.

Levy held the stage in a rosewood-paneled hall of Holland's new Rai Conference Center. Earphones on seats helped overcome the language barrier for the multinational audience. Interpreters in an air-conditioned booth above the hall maintained a running commentary of all addresses and statements. What Levy proposed caused considerable comment.

America, Western Europe, and Japan should get together to form a united front of consumer nations to face the Organization of Petroleum Exporting Countries, he said. He painted an arresting picture of a world being held to ransom by OPEC, with oil companies becoming their junior partners in the situation. The answer is coordination among nations on the consumer side.

"If such a countervailing power to OPEC should really become a factor in international oil, which indeed it must, there is some reasonable hope that international oil and financial arrangements could be set up on a rational and manageable basis; and that OPEC would no longer be able as Sheikh Ahmed Zaki Yamani (of Saudi Arabia) put it in October 1972, through its coordination and unity to 'prove time and time again that it can enforce its demands,' " he told the audience.

But the idea of confronting OPEC with an OPIC (Organization of Petroleum Importing Countries) may be a half-dozen years too late. In London and Paris, foreign ministries show no enthusiasm for the idea. In the Middle East, Sheikh Yamani and others warn that they would view an OPIC as a belligerent act against Arabs.

"I agree strongly that there should be some form of coordination among consumer nations. This should not begin or end with a joint confrontation with OPEC," said Ronald Stuart Ritchie, Senior

Vice-president and Director of Imperial Oil Ltd., Canada's major oil company.

For the United States there is little hope in believing that somehow OPEC's unity may be broken and that divisions in the Arab world then may be fostered for American advantage. Even as Arabs wrangle over political questions, they usually manage to find a basis for cooperation in oil. Arab leaders are sophisticated and wise enough to realize how much they have gained from oil unity. They are unlikely to do anything to diminish OPEC's power.

Oil is in a sellers' market and will remain in one for at least another decade and a half. Already in 1973 Arab oil-producing nations had enough money in their reserves to shut down petroleum production for a year and a half while still maintaining their imports of food and necessities. Storage stocks of oil in the Western world range from only a few days in the United States to about sixty days in Europe, with the latter figure scheduled to rise to ninety days by 1975. In a clash, the need for oil in the industrial world would grow much more acute than would the Arab world's need for cash.

Arabs seemed to hold most of the cards when they introduced production cutbacks in late 1973 to show displeasure at American support for Israel. Europe took the brunt of the oil shortages which developed, along with Japan. Airlines reduced service. Offices cut down on heating. Motorists curtailed their driving. Still, European nations did not want to join through the Organization for Economic Cooperation and Development in any anti-OPEC move, feeling this would be counterproductive.

Events so far do show the seriousness of the world's energy position. The supply-demand picture is so delicately balanced that partnership, not confrontation, is in order. Sheikh Ahmed Zaki Yamani, Saudi Arabia's oil minister, already has signified a willingness to cooperate with the West, and especially with America, to assure continuity of supply. Perhaps, with quiet diplomacy the Israeli time-bomb may be defused, without sacrificing Israel's independence, while assuaging Arab pride. Then, the oil partnership may be drawn.

It is a partnership, however, that perhaps can better be arranged through companies than on a government-to-government basis. Any special government-to-government agreement raises questions about international cooperation, generally. Such a move on America's part could set off a rush of competitive bilateral agreements among other consumer-producer nations. This might not be good for anybody in the consuming area. Major companies have flourished by being as apolitical as possible. Moreover, by 1982 or before, the producing companies will be in partnership with producing states. All companies operating in the Middle East will be owned 51 percent by nations, in those cases where oil-producing countries have not already nationalized them 100 percent. Thus, companies are equipped for cooperating closely with nations.

Close company-producer nation cooperation may ease another formidable problem that faces oil. This is the question of how investments are to be generated to finance the tremendous volume of petroleum that must be delivered to markets over the next decade or so.

One Chase Manhattan Bank study estimates that the petroleum industry's financing requirements total $1 trillion in the period 1970 to 1985. Companies should be able to generate $600 million of that internally, but will have to raise $400 million on financial markets.

"It is doubtful under present economic conditions that the industry will be able to raise that amount," the bank report states.

Both Saudi Arabia and Abu Dhabi have expressed a desire to invest in the downstream operations of oil. Iran already is doing so. When countries have a 51 percent participation, they undertake 51 percent of the investment costs of producing companies. Therefore, through participation and cooperation, the oil industry may be able to solve some of its financing problems, which otherwise might restrict the flow of oil to market.

The political consequences may be delicate, but should not be insurmountable. Who would have thought a few years ago that American companies would be discussing joint ventures with the Soviet Union in the oil and gas area? Two agreements alone call for American investments of $10 to $20 billion in Siberian natural-gas fields. American companies, in turn, will receive natural gas for transport in tankers to the United States, where it would be sold.

Israel is a political problem of another gender. Ironically, it seems easier to bury American-Russian differences than to settle the Arab-Israeli question. Yet it is one that is being thrust into the foreground by energy matters.

In 1972 in testimony presented to the Platform Committee of the Democratic National Committee, John P. Richardson, speaking on behalf of the Middle East Affairs Council, spoke on this matter:

> The Middle East Affairs Council maintains that it is possible to insure the security of Israel without necessitating the utter insecurity of all its neighbors. It is possible to demonstrate concern for Jewish lives in Israel without having to be silent about the Palestinian Arabs, who have been made a people without a land. It is possible to accommodate domestic political interests without seeing our country driven out of the Arab World altogether.

Finding a solution to the Arab-Israeli problem in the Middle East will not be easy. Nobody with any sense is suggesting that Israel be abandoned. In fact, that wouldn't produce one extra barrel of oil in a tight situation. But America's oil crisis does intensify the need to find a peaceful solution. Like it or not, the United States finds it necessary to deal with the Middle Easterner on a friendly and cooperative basis. Hopefully, long before the 1980s arrive, the term "Middle Easterner" will include Arabs, Iranians, Jews, and minorities, all living together in peace as the profits from oil seep through the entire region.

INDEX

INDEX